Automated Structural Analysis:
An Introduction

Pergamon Unified
Engineering Series

**Pergamon Unified
Engineering Series**

Automated Structural Analysis: An Introduction

Professor William R. Spillers
Dept. of Civil Engineering & Engineering Mechanics,
Columbia University

PERGAMON PRESS INC.,
NEW YORK · TORONTO · OXFORD
SYDNEY · BRAUNSCHWEIG

Pergamon Press Inc., Maxwell House, Fairview Park, Elmsford, New York 10523

Pergamon of Canada Ltd., 207 Queen's Quay West, Toronto 1

Pergamon Press Ltd., Headington Hill Hall, Oxford

Pergamon Press (Aust.) Pty. Ltd., 19a Boundary Street, Rushcutters Bay, N.S.W. 2011 Australia

Vieweg & Sohn GmbH, Burgplatz 1, Braunschweig

Printed in the United States of America
08 016782 9(H)

Pergamon
Unified Engineering
Series

GENERAL EDITORS

Thomas F. Irvine, Jr.
State University of New York at Stony Brook
James P. Hartnett
University of Illinois at Chicago Circle

EDITORS

William F. Hughes
Carnegie-Mellon University
Arthur T. Murphy
PMC Colleges
William H. Davenport
Harvey Mudd College
Daniel Rosenthal
University of California, Los Angeles

SECTIONS

Continuous Media Section
Engineering Design Section
Engineering Systems Section
Humanities and Social Sciences Section
Information Dynamics Section
Materials Engineering Section
Engineering Laboratory Section

Contents

Contents

Preface

The typical university undergraduate structural analysis sequence has for some time consisted of a course in statically determinate structures followed by a course on indeterminate structures. While statically determinate structures are certainly important for many practical reasons, the logic of this sequence lay in the fact that statically determinate structures played a central role in the methods used for the analysis of statically indeterminate structures.

This is no longer the case. Today the common method of analysis is the node method – a method in which the concept of statical indeterminacy plays a minor role. Beyond that, since the advent of the computer it has become productive to think about structures in more general terms. These arguments and many others, including the recent interest in finite element methods, have motivated this book.

It would be fatuous to claim that this text offers a complete coverage of structural analysis for the undergraduate – nor is that the intent. Roughly, this text attempts to cover half of an undergraduate program.

There are essentially two requirements placed upon a structural analysis undergraduate sequence today: (1) the student must learn the formal aspects of structural analysis and (2) he must be able to deal with the analysis problems which are associated with design and which are frequently handled in an approximate manner. It is the thrust of this text to separate the formal aspects of analysis and treat them separately from the design oriented, back of the envelope techniques with which they have so little in common. The payoff for this separation comes not in the beauty of an abstract presentation, but in its utility; purely and simply, the more abstract presentation is almost trivial to program.

Perhaps it is more reasonable to consider the entire undergraduate structures sequence. Certainly design must be included and there is always some requisite analysis associated with the teaching of design. The real question is what to do about analysis itself. With the exception of some rather simple classes of problems, analysis today is done by computer; if meaningful problem solving is to be included in an undergraduate structures program it must be done on the computer, and it is here that the classical methods present difficulties. The difficulties of automating the classical methods have led to the development of new methods of analysis; this text attempts to introduce these methods on an elementary level.

With regard to background, this text requires relatively little of the student. Taken on its lowest level, it can be used to discuss structures composed of straight uniform members or arbitrary members as they can be approximated by straight uniform members. At this level the student need only know the behavior of rods and straight beams from solid mechanics. In general it is presumed that the student is comfortable with matrix notation and has a working knowledge of FORTRAN.

Finally, this book is the product of some 12 years' undergraduate teaching by its author at Columbia University. While it would surely be impossible to identify and give credit to all those whose influence is reflected here, it should be noted that this book was conceived by Professor Frank DiMaggio and its author sometime in 1962. The fact that it exists in its present form can be attributed to the encouragement of Professor Mario Salvadori (a delightful spirit), the indirect support of the National Science Foundation and Dr. Michael Gaus, director of their Engineering Mechanics program, without which this text would have remained in its most rudimentary manuscript form, and the suggestion by Professor W. F. Hughes, a Pergamon editor, that it be included in their Unified Engineering Series.

Dept. of Civil Engineering   William R. Spillers
  & Engineering Mechanics
Columbia University

ONE

Graphs, Networks, and Structures

1.1 INTRODUCTION

In modern automated structural analysis it is the practice to emphasize generality with the result that computer programs rarely deal with a specific structure but rather with classes of structures. It is this practice which motivates proceeding from truss, to plane frame, to space frame — classes of problems with increasing complexity. This first chapter however attempts to present in a rather descriptive manner the ideas or laws which are fundamental to structures. These ideas are most simply presented for the network, a degenerate class of structures if you like.

Probably the second most significant development of recent years has been the separation of what is called the "connectivity" of a structure from the description of the elements from which the structure is being assembled. Basically it has become the practice to separate the discussion of the mechanical properties of the elements themselves from the discussion of the manner in which they are connected. The latter is concerned simply with points which may or may not be connected by lines, is independent of geometric quantities such as distance or spatial relationships, and is called graph theory. It is with the elements of graph theory that this text begins formally in the next section.

Beyond this, there is a unity among physical systems which has been exploited recently in various attempts at a unified system theory; certainly from a pedagogical point of view it is important to understand the relationships which are common to electrical, mechanical, and structural systems. Nowhere do these relationships stand out as clearly as they do in the system graph. This fact alone might motivate beginning a book on structures with an introduction to graph theory.

There is, however, another more practical reason for doing so. Of the systems commonly encountered, structures are by far the most complex and for that reason it is important to make use of any heuristics which may be available. In particular, it can be useful to think of the methods of structural analysis as generalizations of the methods of network analysis which have stronger intuitive appeal and are in fact quite simple.

Finally, structures research commonly makes sophisticated use of graph theoretical techniques which are easily introduced at an early level. In any case, it is the utility of graph theory, not its elegance, which motivates its inclusion in this text.

1.2 GRAPH THEORY

Fundamental to the organization of this text is the idea of assemblying pieces to form a structure. The properties of the composite structure are then obtained from the properties of the pieces of which it is composed and depend upon the manner in which the pieces are put together — both the geometric properties such as the angles with which the members intersect and the topology or connectivity of the structure which describes the points at which, e.g., element *b* connects with element *c*. While the approach presented in this text is valid for arbitrary elements, this book is concerned in detail only with the assembly of elements (rods and beams) which can be represented by lines. For a discussion of more complex elements (plates, shells, etc.) the reader is referred to the interesting new book of Azar or the comprehensive treatment of Zienkiewicz. In this section some of the elements of graph theory are presented. The reader will note a remarkable utility of graph theory in network analysis as it is described in the section which follows; this same utility, while demanding more effort in its application, persists throughout all structures.

The first element encountered in graph theory is the *node* or point. In structures this may correspond to a physical joint, an arbitrary reference point in an otherwise continuous member, or some other quantity which it is convenient to represent as a node. Just as two joints of a structure may be connected by a member, two nodes may be connected by a *branch* or line. A *graph* then is defined by listing a set of nodes, X, and a set of branches, U (which may be described by listing node pairs). For the graph in Fig. 1.1 the set of nodes is

$$X = (1, 2, 3, 4, 5, 6, 7, 8),$$

and the set of branches

$$U = [(2, 1), (3, 2), (7, 3), (5, 4), (6, 5), (8, 6), (1, 4), (2, 5), (3, 6), (7, 8)].$$

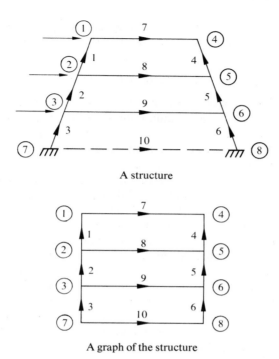

A structure

A graph of the structure

Fig. 1.1.

(A fictitious branch has been added to the structure to represent the support offered it by the ground. This step will be formalized later in the text.)

When it is necessary to distinguish between the two nodes of a given branch, it can be done by agreeing that when a branch is described by a node-pair, the first node listed is to be referred to as the negative node and the second the positive node, as shown in Fig. 1.2. It is convenient to augment this description by an arrow placed on the branch which indicates a direction from its negative to its positive end. When this is done the graph is called a *directed graph*.

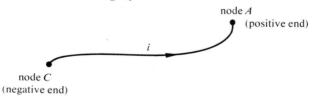

Fig. 1.2. The *i*th branch.

It is now possible to describe a *path* from say node *a* to node *b* by listing a sequence of nodes and branches. In Fig. 1.1 one path from node 1 to node 7 is

$$1, (2,1), 2, (2,5), 5, (6,5), 6, (3,6), 3, (7,3), 7.$$

Having the definition of a path, it is a simple matter to describe a *mesh* (closed loop) as a path whose initial and final nodes are identical and in which only the initial node is repeated. In a *connected graph* there exists at least one path between each node in the graph. A *tree* is simply a connected graph which contains no meshes.

Graph theory is concerned with the description and exploitation of the "connectivity" (i.e. the information describing which branches are connected to which nodes) of a graph and its implications. It is a very amorphous discipline with a wide range of applications. In the remainder of this section several matrices are described which are directly useful in network theory. To the reader who wishes to pursue graph theory are available the new book of Maxwell and Reed, the classics of Berge and Ore and the very delightful network oriented paper of Branin.

For the purposes of this book the most useful description of the connectivity of a graph or structure follows from the *augmented branch node incidence matrix $\bar{A}$*. Given a graph which contains *n* nodes and *b* branches, the matrix $\bar{A}$ is defined to be a $b \times n$ (row $\times$ column) matrix whose elements $\bar{A}_{ij}$ are

$$\bar{A}_{ij} = \begin{cases} +1 & \text{if node } j \text{ is the positive end of member } i \\ -1 & \text{if node } j \text{ is the negative end of member } i \\ 0 & \text{otherwise} \end{cases}$$

For the graph shown in Fig. 1.1 this matrix is

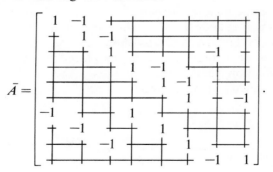

The *branch-node incidence matrix A* is obtained from $\bar{A}$ by deleting certain reference nodes such as support nodes in structures. The matrix *A*

is of particular importance to the node method (*see* Chapter 3) and generalizations of it appear repeatedly throughout this text.

A second description of the connectivity of a graph is contained in the *branch-mesh matrix* C. For a graph with b branches and m meshes, C is a $b \times m$ (row $\times$ column) matrix whose elements are

$$C_{ij} = \begin{cases} +1 & \text{if branch } i \text{ is positively oriented in mesh } j \\ -1 & \text{if branch } i \text{ is negatively oriented in mesh } j \\ 0 & \text{otherwise} \end{cases}$$

In order to determine whether a branch is positively or negatively oriented in a mesh it is only necessary to see whether or not the direction of the branch agrees with the arbitrarily assumed direction of the mesh. While this procedure is intuitive, it can be formalized as follows: Given a connected graph, delete branches until the graph becomes a tree. The branches which have been deleted are called *links*; each link defines a mesh since its insertion into the tree forms a closed path; these meshes are "independent" since each link occurs only in one mesh; if the direction of the mesh is assumed to coincide with the direction of the link which defines it, the orientation of a branch in the mesh is positive or negative depending upon whether positive motion around the mesh agrees with or opposes the direction of the branch.

For the graph in Fig. 1.1 one possible branch-mesh matrix is

$$C = \begin{bmatrix} -1 & -1 & -1 \\ + & -1 & -1 \\ + & + & -1 \\ 1 & 1 & 1 \\ + & 1 & 1 \\ + & + & 1 \\ -1 & -1 & -1 \\ 1 & + & + \\ + & 1 & + \\ + & + & 1 \end{bmatrix}.$$

Note that as the example indicates, it is always possible to write C in the form

$$C = \left[\frac{C_T}{I} \right],$$

in which I is the identity matrix although later in the book examples will indicate that this form may not be efficient.

While the construction of the branch mesh matrix, given a graph, is not a unique process, the number of meshes in a given graph is invariant.

To demonstrate this, note first that a tree with n nodes contains $n-1$ branches. (This follows from the fact that a tree can be constructed by starting with two nodes and a single branch and adding, a step at a time, one branch and one node until the tree is complete.) It follows directly that a connected graph with n nodes and b branches contains

$$m = b - (n-1) = b - n + 1,$$

meshes and links. The number m is clearly independent of the manner in which the meshes are constructed.

The matrices A and C satisfy the relationship

$$\tilde{A}C = 0. \tag{1.1}$$

Proof: In terms of its elements, Eq. (1.1) is

$$\sum_{k=1}^{b} A_{ki}C_{kj} = 0. \tag{1.2}$$

Since each column of A describes the branches incident upon a given node and since only two of these branches lie in any mesh which includes the node, there are two non-zero terms in each of Eq. (1.2) if there are any. Figure 1.3 shows the three possible cases, each of which agrees with the theorem.

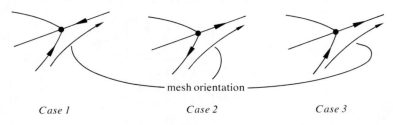

— mesh orientation —

Case 1   *Case 2*   *Case 3*

Fig. 1.3.

This section will be concluded with a definition of the cut-set matrix. First note that given a connected graph, it is always possible to construct a tree which contains all the nodes of the original graph simply by deleting branches (a set of links). Since removing any branch from a tree results in a disconnected graph, each tree branch divides the nodes into two sets. A *cut-set* can now be defined to be a set of branches whose removal from a graph results in dividing the graph into two unconnected graphs. Clearly, one way to construct a cut-set is to select one tree branch (which defines the two sets of nodes) and each link which has one node in each of these node sets.

For a connected graph with n nodes, b branches, and m meshes, the number of tree branches, t is just $b-m$. The cut-set matrix D, is then defined to be a $b \times t$ matrix whose elements are

$$D_{ij} = \begin{cases} +1 & \text{if branch } i \text{ is positively oriented in the } j\text{th cut-set} \\ -1 & \text{if branch } i \text{ is negatively oriented in the } j\text{th cut-set} \\ 0 & \text{otherwise} \end{cases}$$

As in the case of the branch-mesh matrix, the branch which defines a cut-set can be used to define the orientation of the cut-set. That is, all members of the cut-set which have their positive node in the set which contains the positive node of the defining tree branch are said to be positively oriented in the cut-set while the others are said to be negatively oriented. For the example shown in Fig. 1.1 a cut-set matrix is

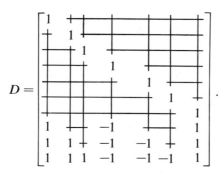

1.3 PIPING NETWORKS

Network theory in general and piping networks in particular provide an immediate, simple application of the aspects of graph theory which were developed in Section 1.2. Since this section is peripheral to the text, only the most elementary piping problem is considered: Given a system of pipes whose properties (length, diameter, and roughness) are known and from whose nodes known quantities of fluid per unit time are being drawn; find the flows and head losses in each of the pipes (branches) and the pressure at each of the nodes. Since their descriptions do not involve the geometric locations of nodes, problems of this kind are called "topological network problems".

Whatever the flows and pressures in the network, they must satisfy the

NODE LAW. *The sum of the flows into any node must equal the sum of the flows out of the node.*

7

and the

MESH LAW.   *The sum of the head losses around any mesh must equal zero.*

In addition to the node and mesh laws, the network solution must satisfy a "constitutive equation" which relates the flow in each pipe (branch) to its head loss. With the help of the matrices defined in Section 1.2 it is now possible to make a precise statement of this network problem.

Associated with the ith node of the network are a pressure δ_i and a flow being drawn out of the system at that node P_i. Since the flow into the network as a whole must equal the flow out of the network (there are no places where fluid can collect)

$$\sum_{i=1}^{n} P_i = 0, \tag{1.3}$$

in which n is the total number of nodes. Since the problem statement only involves "head losses", pressures are only determined to within an additive constant which allows the pressure at node n to be arbitrarily set to zero. Node n then becomes the "reference node" for the network.

Associated with the ith branch (pipe) of the network are a flow F_i and a head loss Δ_i (*see* Fig. 1.4). The branch quantities are related by the non-linear constitutive equation

$$F_i = K_i(\Delta_i)^\alpha, \qquad (i = 1, \ldots, b) \tag{1.4}$$

in which the exponent α is given. It is convenient at this point to assemble these physical variables into the matrices,

δ — node pressure matrix,
P — node flow matrix,
Δ — branch pressure (head-loss) matrix,
F — branch flow matrix,
K — branch roughness matrix.

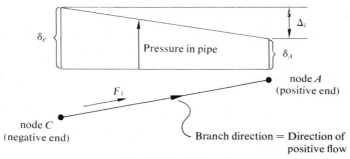

Fig. 1.4. The ith branch.

Which are simply

$$\delta = \begin{bmatrix} \delta_1 \\ \delta_2 \\ \cdot \\ \cdot \\ \cdot \\ \delta_{n-1} \end{bmatrix}, \qquad P = \begin{bmatrix} P_1 \\ P_2 \\ \cdot \\ \cdot \\ \cdot \\ P_{n-1} \end{bmatrix},$$

and

$$\Delta = \begin{bmatrix} \Delta_1 \\ \Delta_2 \\ \cdot \\ \cdot \\ \cdot \\ \Delta_b \end{bmatrix}, \qquad F = \begin{bmatrix} F_1 \\ F_2 \\ \cdot \\ \cdot \\ \cdot \\ F_b \end{bmatrix}, \qquad K = \begin{bmatrix} K_1 & & & & \\ & K_2 & & \bigcirc & \\ & & \cdot & & \\ & \bigcirc & & \cdot & \\ & & & & K_b \end{bmatrix},$$

where b is the number of branches in the system.

In Fig. 1.5 the graph of Fig. 1.1 is now treated as a piping system while Fig. 1.6 shows a free body diagram of node 6. For this node, the node law requires that

$$F_9 + F_6 - F_5 = P_6. \tag{1.5}$$

In general, if a branch (pipe) is positively (negatively, not) incident on a node, its flow appears positively (negatively, does not appear) in that node equation. In view of the definition of the branch node incidence matrix, the node equations for the system are

$$\tilde{A}F = P. \tag{1.6}$$

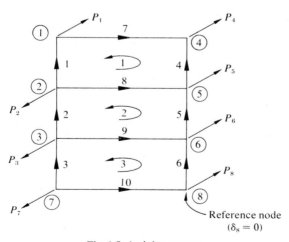

Fig. 1.5. A piping system.

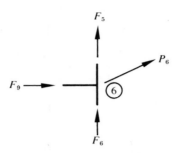

Fig. 1.6. A free body diagram of node 6.

For piping systems there is a single reference node so that the matrix A is $b \times (n-1)$. The matrix A will be returned to shortly.

In terms of the matrices defined, the constitutive equations are just

$$F = K\Delta^{\alpha}. \tag{1.7}$$

In a piping system it is most common to work with positive pressures which tend to compress the fluid. This is contrary to the convention used in solid mechanics in which a positive mean pressure tends to produce an increase in volume. Figure 1.4 indicates the usual convention showing the head loss–node pressure relationship

$$\Delta_i = \delta_c - \delta_A. \tag{1.8}$$

In general, to find the head loss along a branch it is only necessary to subtract the pressure at the positive end from the pressure at the negative end of the branch. Again referring to the definition of the branch node matrix, this relationship can be written for the entire system as

$$\Delta = -A\delta. \tag{1.9}$$

Returning to the matrix A, recall that it was derived from the augmented branch node incidence matrix by deleting a reference node. It is now clear that the deletion of a reference node plays a dual role. In view of Eq. (1.3), the node equations are not independent and the deletion of a reference node has the effect in Eq. (1.6) of eliminating a redundant equation. With regard to Eq. (1.9), the deletion of the reference node has the effect of forcing its pressure to be zero. It will be seen later that these remarks all have implications in structures.

It is now possible to give a formal statement of the *node method* for piping systems which follows directly after collecting Eqs. (1.6, 1.7, and 1.9)

$$\tilde{A}F = P \qquad \text{(node equation)}$$
$$F = K\Delta^\alpha \qquad \text{(constitutive equation)}$$
$$\Delta = -A\delta \qquad \text{(branch pressure-node pressure equation)}$$

Then

$$\tilde{A}F = P \quad \rightarrow \quad \tilde{A}K\Delta^\alpha = P \quad \rightarrow \quad \tilde{A}K(-A\delta)^\alpha = P. \qquad (1.10)$$

Equation (1.10) is a system of nonlinear simultaneous equations on the unknown node pressures δ.

A more common formulation for piping systems is the *mesh method*. In order to be able to present this method in a formal manner it is first necessary to state the mesh law precisely. Briefly, the idea is that you go around a mesh, the head losses of each branch when given the proper sign must sum to zero in order to be able to return to the pressure at the starting node. For mesh 2 in Fig. 1.5 the mesh law gives

$$-(\delta_2 - \delta_5) + (\delta_6 - \delta_5) + (\delta_3 - \delta_6) - (\delta_3 - \delta_2) = 0$$

or

$$-\Delta_8 + \Delta_5 + \Delta_9 - \Delta_2 = 0. \qquad (1.11)$$

Generalizing a little and using the definition of head loss given in Eq. (1.8), the mesh equation for a given mesh contains a head loss term for every branch in the mesh. The term carries a positive or negative sign depending upon whether the branch is positively or negatively oriented in the mesh in question. In view of the definition of the branch mesh matrix, it follows directly that the mesh law for the entire network can be written

$$\tilde{C}\Delta = 0. \qquad (1.12)$$

Since each link defines a mesh, it is a simple matter to use the links to define flows which satisfy the node law for the case in which no fluid is being drawn from the system — flows in which fluid just circulates around the mesh. Such a flow for mesh 2 of the system in Fig. 1.5 is

$$F_8 = -1 \qquad F_5 = +1 \qquad F_9 = +1 \qquad F_2 = -1,$$

with all other branch flows zero and $P = 0$.

A convenient way to describe this flow is to say that the flow in mesh 2 is equal to 1. To generalize the description of mesh flows it is useful

11

to partition the branch flow matrix into tree branch flows F_T and link flows F_L,

$$\begin{bmatrix} F_1 \\ F_2 \\ . \\ . \\ . \\ F_b \end{bmatrix} = \begin{bmatrix} F_T \\ F_L \end{bmatrix}. \tag{1.13}$$

(As a matter of convenience here the links are numbered last in the matrix F.) Returning again to the definition of the branch mesh matrix, mesh flows can be written for the entire system as

$$F = CF_L, \tag{1.14}$$

which gives the example just described when

$$F_L = \begin{bmatrix} 0 \\ 1 \\ 0 \end{bmatrix}. \tag{1.15}$$

However, the mesh flows themselves do not satisfy the node law when $P \neq 0$. In this case it is necessary to add any known flow $\bar{F}$ with the property

$$\tilde{A}\bar{F} = P \tag{1.16}$$

to Eq. (1.14) to obtain the general form of the branch flows

$$F = \bar{F} + CF_L \tag{1.17}$$

which clearly satisfies the node law $\tilde{A}F = P$ since $\tilde{A}\bar{F} = P$ and $\tilde{A}C = 0$.

While $\bar{F}$ can be found in many ways, one simple procedure is to set the link flows to zero and proceed inward from the tips of the tree branches computing flows using the node law. For the example in Fig. 1.5 this leads to

$$\bar{F} = \begin{bmatrix} -P_7 & -P_3 & -P_2 \\ -P_7 & -P_3 \\ -P_7 \\ -P_8 & -P_6 & -P_5 \\ -P_8 & -P_6 \\ -P_8 \\ P_8 & +P_6 & +P_5 & +P_4 \\ 0 \\ 0 \\ 0 \end{bmatrix}.$$

The *mesh method* follows directly as

$$\tilde{C}\Delta = 0 \qquad \text{(mesh equation)}$$
$$\Delta = K^{1/\alpha}F^{1/\alpha} \qquad \text{(constitutive equation)}$$
$$F = \bar{F} + CF_L \qquad \text{(branch flow-mesh flow equation)}$$

$$\tilde{C}\Delta = 0 \quad \rightarrow \quad \tilde{C}K^{1/\alpha}F^{1/\alpha} = 0 \quad \rightarrow \quad \tilde{C}K^{1/\alpha}(\bar{F}+CF_L)^{1/\alpha} = 0 \qquad (1.18)$$

Equation (1.18) is a system of nonlinear simultaneous algebraic equations on the unknown mesh flows F_L which may be solved iteratively.

1.3.1 An Example

It is difficult to find a meaningful piping problem for solution by hand computation because the nonlinear constitutive equations result in nonlinear algebraic equations for both the node and mesh method and systems of nonlinear algebraic equations are an order of magnitude more difficult than linear systems – the primary concern of this text. In the example used in this section, the network formulation is linearized by assuming $\alpha = 1$ which does not seem to correspond to any realistic water supply system but is in the spirit of linear structural analysis.

Figure 1.7 shows a simple piping system which will first be solved using the node method and then, using the mesh method. In this case the relevant matrices are

$$F = \begin{bmatrix} F_1 \\ F_2 \\ F_3 \\ F_4 \end{bmatrix} \qquad \Delta = \begin{bmatrix} \Delta_1 \\ \Delta_2 \\ \Delta_3 \\ \Delta_4 \end{bmatrix} \qquad P = \begin{bmatrix} P_1 \\ P_2 \end{bmatrix} = \begin{bmatrix} 1 \\ 1 \end{bmatrix} \qquad \delta = \begin{bmatrix} \delta_1 \\ \delta_2 \end{bmatrix}$$

$$A = \begin{bmatrix} 1 & -1 \\ 0 & 1 \\ 1 & -1 \\ 0 & 1 \end{bmatrix} \qquad C = \begin{bmatrix} -1 & 0 \\ 0 & -1 \\ 1 & 0 \\ 0 & 1 \end{bmatrix}$$

and for simplicity it is assumed that K has the form

$$K = K_0 \begin{bmatrix} 1 & & & \\ & 2 & & \\ & & 1 & \\ & & & 2 \end{bmatrix}$$

in which K_0 is an arbitrary positive constant.

The node method, Eq. (1.10), requires the solution of

$$-(\tilde{A}KA)\delta = P$$

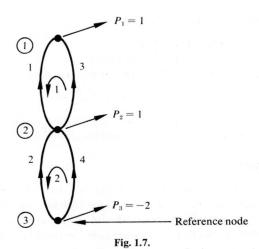

Fig. 1.7.

when $\alpha = 1$. In this case the system to be solved is

$$-\begin{bmatrix} 1 & 0 & 1 & 0 \\ -1 & 1 & -1 & 1 \end{bmatrix} K_0 \begin{bmatrix} 1 & & & \\ & 2 & & \\ & & 1 & \\ & & & 2 \end{bmatrix} \begin{bmatrix} 1 & -1 \\ 0 & 1 \\ 1 & -1 \\ 0 & 1 \end{bmatrix} \begin{bmatrix} \delta_1 \\ \delta_2 \end{bmatrix} = \begin{bmatrix} 1 \\ 1 \end{bmatrix},$$

or

$$-K_0 \begin{bmatrix} 2 & -2 \\ -2 & 6 \end{bmatrix} \begin{bmatrix} \delta_1 \\ \delta_2 \end{bmatrix} = \begin{bmatrix} 1 \\ 1 \end{bmatrix},$$

from which

$$\delta = \frac{-1}{K_0} \begin{bmatrix} 1 \\ \frac{1}{2} \end{bmatrix},$$

and

$$\Delta = -A\delta = \frac{1}{K_0} \begin{bmatrix} \frac{1}{2} \\ \frac{1}{2} \\ \frac{1}{2} \\ \frac{1}{2} \end{bmatrix} \qquad F = K\Delta = \begin{bmatrix} \frac{1}{2} \\ 1 \\ \frac{1}{2} \\ 1 \end{bmatrix}.$$

Note the predictable symmetry here.

Using the mesh method to re-solve this problem requires the solution of Eq. (1.18),

$$(\tilde{C}K^{-1}C)F_L = -\tilde{C}K^{-1}\bar{F}.$$

Partitioning the F matrix as indicated earlier results in

$$F_L = \begin{bmatrix} F_3 \\ F_4 \end{bmatrix}.$$

Setting $F_L = 0$, $\bar{F}$ become (by inspection)

$$\bar{F} = \begin{bmatrix} 1 \\ 2 \\ 0 \\ 0 \end{bmatrix}.$$

The system to be solved is then

$$\begin{bmatrix} -1 & 0 & 1 & 0 \\ & & & \\ 0 & -1 & 0 & 1 \end{bmatrix} K_0^{-1} \begin{bmatrix} 1 & & & \\ & \frac{1}{2} & & \\ & & 1 & \\ & & & \frac{1}{2} \end{bmatrix} \begin{bmatrix} -1 & 0 \\ 0 & -1 \\ 1 & 0 \\ 0 & 1 \end{bmatrix} \begin{bmatrix} F_3 \\ F_4 \end{bmatrix}$$

$$= -\begin{bmatrix} -1 & 0 & 1 & 0 \\ & & & \\ 0 & -1 & 0 & 1 \end{bmatrix} K_0^{-1} \begin{bmatrix} 1 & & & \\ & \frac{1}{2} & & \\ & & 1 & \\ & & & \frac{1}{2} \end{bmatrix} \begin{bmatrix} 1 \\ 2 \\ 0 \\ 0 \end{bmatrix}$$

or

$$\begin{bmatrix} 2 & 0 \\ 0 & 1 \end{bmatrix} \begin{bmatrix} F_3 \\ F_4 \end{bmatrix} = \begin{bmatrix} 1 \\ 1 \end{bmatrix},$$

from which

$$F_L = \begin{bmatrix} F_3 \\ F_4 \end{bmatrix} = \begin{bmatrix} \frac{1}{2} \\ 1 \end{bmatrix},$$

and

$$F = \bar{F} + CF_L = \begin{bmatrix} \frac{1}{2} \\ 1 \\ \frac{1}{2} \\ 1 \end{bmatrix}.$$

$\Delta = K^{-1}F$ can now be computed to complete this example.

1.4   SKELETAL STRUCTURES

1.4.1 Types of Structures Considered

Fundamental to this section is the concept of "modeling". That is, when given a real structure to analyze, it is always necessary to approximate the real structure, which is much too complex to solve, by some "model" of it which is simple enough to work with. It is in the modeling that the physics of a problem appears; certainly modeling is the most decisive step in structural analysis and probably the most difficult. It only becomes trivial for classes of problems which have through many years' experience become well understood.

In structural analysis, a structure and some sort of "loading" are given from which the "response" is determined. The structure can be an aircraft, a building, a submarine; the loading can be the effect of gravity, wind, heating, earthquake; the response is usually the displacements and internal stress resultants. The engineer must take the structure and decide upon a suitable method of analysis.

The first step in analysis is to classify the structure. While the available classifications are vague, for the purposes of this book it is only necessary to decide whether or not the structure is "skeletal" (i.e. can be described by lines and quantities associated with lines) since only skeletal structures are dealt with in any detail here. In some cases this decision is a simple matter. Some structures have a clearly defined "structural frame" through which the loads are transferred. The most common example of this is the multistory steel building which has a clearly defined structural frame once the curtain walls and floor slabs are removed; another example is the slow speed aircraft which also has a visible structural frame once the aerodynamic covering has been removed. In other cases such as high speed aircraft wings or structural shells, the load carrying part of the structure may clearly not be a frame (i.e. made up of linear elements) but even in these cases it may be useful to approximate the actual structure by a skeletal structure. Certainly there are truss and frame models of shells.

Having identified the structural skeleton, the second step is to model the elements of which it is composed. Since the common structural elements are discussed at length in the subsequent chapters of this book, only some of the more general restrictions will be noted here. They are all wrapped up neatly in the term "linear structure" which means that the model used is described by linear equations, usually a system of simultaneous linear, algebraic equations. This implies that:

1. The elements are elastic. Since perfect elasticity is an ideal not achieved in nature, this means that the situation into which the structure has been placed is simple enough to allow the real structural elements to be modeled by "generalized springs" for which generalized *linear* force-displacement relationships can be written.

2. The structure undergoes small displacements. While the details of the small displacement restriction (which usually only means small rotations) are also discussed in subsequent chapters, it may be noted here that the implication is always that the geometry of the structure is the same after loading as it was before.

There are many, many more remarks which could be made concerning modeling; they would certainly include the effects of the depth of members, connections, material homogeneity, etc. However they generally require sophistication and experience beyond the scope of an elementary text.

1.4.2 Requirements for a Solution

The methods described in this book are usually referred to as "finite element" methods. This simply means that the emphasis is placed upon decomposing a structure into elements or pieces which are re-assembled so that their forces and associated displacements satisfy certain requirements.

Quite informally, there are three requirements which are made on any solution in linear structural analysis:

1. Equilibrium must be satisfied.
2. Hooke's law (i.e. the prescribed linearity between generalized spring forces and generalized spring displacements) must be maintained.
3. The pieces of the structure must fit together when they are assembled in their deformed state.

This book is concerned with describing methods for obtaining solutions which satisfy these requirements. Historically, many approaches have been taken — some of which are outlined in Appendix 2 and referred to as classical methods. Fortunately, one method, the node method predominates in (computer) applications today; this method is introduced in the next chapter.

TWO

The Node Method for Trusses

2.1 INTRODUCTION

It is the function of this chapter to provide a simple, concrete, structural example of the node method from which the reader is asked to generalize in Chapter 3. This sequence is, of course, based upon the idea that it is easier to generalize upward from a simple example (at least in an elementary course) than it is to accept a theory in terms of which all applications are special cases. Those who prefer the general case first will find it a simple matter to interchange for themselves the order of Chapters 2 and 3.

Because of its simplicity, the truss provides an ideal introduction to the node method, although it will be seen later that in the case of the mesh method, the truss is in at least one respect (due to the existence of releases), more difficult than the frame. In this chapter a direct approach to the truss problem is used. This requires first a formal description of the forces and displacements which occur, then a translation of the structural equations into matrix form. It is the form of this translation which is motivated by the network problem and which results in a generalization of the branch node incidence matrix.

2.2 THE TRUSS

In the context of this book, the truss has a very restricted definition: It is a structure composed of pin-connected elements which is loaded only at its joints. The implication of pin-connected members which are not loaded along their length is well known from elementary mechanics, and is simply that the resultant member force lies along a straight line between the ends of the member. Since the line of action of the member force is

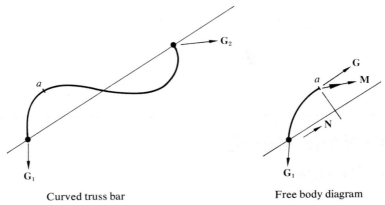

Curved truss bar   Free body diagram

Fig. 2.1.

known, it can be described by specifying a single scalar, the bar force. Figure 2.1 shows a general curved truss bar and then a free body diagram of a part of it. The curved bar can be recognized as a truss bar because *only* forces, no moments, are shown to be acting at its ends. (By definition, a pin-ended member can have no applied end moments.) Since the sum of the forces acting on the member as a whole must be zero,

$$\mathbf{G_1} + \mathbf{G_2} = 0 \quad \text{or} \quad \mathbf{G_1} = -\mathbf{G_2}.$$

Let $\mathbf{r}$ be the position vector of the upper end with respect to the lower end of the member. Taking moments about the lower end,

$$\mathbf{r} \times \mathbf{G_2} = 0$$

from which it follows that since $|\mathbf{r}| \neq 0$, when $\mathbf{G_2} \neq 0$, $\mathbf{G_2}$ is parallel to $\mathbf{r}$. The vector $\mathbf{G_2}$ is then a vector whose line of action is known and which can therefore be described as a scalar multiplying either the vector $\mathbf{r}$ or a unit vector $\mathbf{n}$. Figure 2.2 shows a structure composed of pin-connected bars subjected to joint loads.

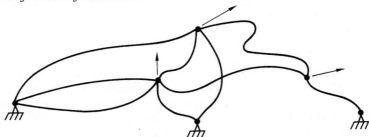

Fig. 2.2. A typical truss.

2.3 A FORMAL DESCRIPTION OF THE TRUSS PROBLEM

The remainder of this chapter is concerned with a rather simple truss problem, the response of a structure composed of *straight*, pin-connected members subjected only to joint loads. While curved truss members are sometimes useful when, e.g., connections introduce eccentricity, they are not treated in the remainder of this book (but only require an appropriate modification of the "primitive stiffness matrix" described later). Such effects as thermal loads, support settlement, lack of fit are added in a later chapter which shows them to be "secondary" with regard to the problem solution.

In truss analysis members (branches) are idealized as lines which meet at points (nodes) which are called joints. Figure 2.3 shows a typical truss joint and the quantities associated with it,

P_i — the applied joint load vector at joint i
δ_i — the displacement vector at joint i

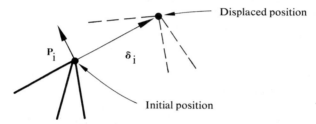

Fig. 2.3. The ith point.

For the entire structure these quantities are assembled into the joint load matrix P and the joint displacement matrix δ,

$$\delta = \begin{bmatrix} \delta_1 \\ \delta_2 \\ \cdot \\ \cdot \\ \cdot \\ \delta_J \end{bmatrix} \quad \text{and} \quad P = \begin{bmatrix} P_1 \\ P_2 \\ \cdot \\ \cdot \\ P_J \end{bmatrix},$$

in which

$$\delta_i = \begin{bmatrix} (\delta_i)_x \\ (\delta_i)_y \\ (\delta_i)_z \end{bmatrix} \quad \text{and} \quad P_i = \begin{bmatrix} (P_i)_x \\ (P_i)_y \\ (P_i)_z \end{bmatrix}.$$

The matrices δ_i and P_i simply contain the components of the vectors δ_i and P_i respectively; J is the number of movable joints (i.e. joints which are not supports).

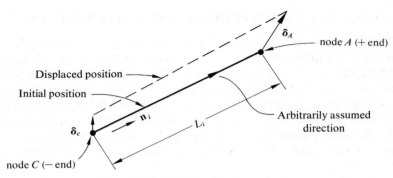

Fig. 2.4. Member i.

Figure 2.4 shows a typical truss bar associated with which is the bar force F_i and the bar length change Δ_i, chosen so that positive F_i and Δ_i correspond to tension or stretching within the bar and a unit vector $\mathbf{n}_i$. Given the displacements of the ends of a bar, it is possible to compute the exact length change of the bar. In linear structural analysis it is customary to use the approximate relationship,

$$\Delta_i = \mathbf{n}_i \cdot (\boldsymbol{\delta}_A - \boldsymbol{\delta}_C) = \tilde{n}_i (\delta_A - \delta_C), \tag{2.1}$$

which uses the notation of Fig. 2.4. This involves projecting the joint displacement vectors onto the original position of the bar to obtain the length change. It is only exact when the displaced position of the bar is parallel to its original position but is accurate when they differ by a *small rotation*. In Eq. (2.1) $\mathbf{n}_i$ is a unit vector which describes the slope of bar i and which has the matrix representation

$$n_i = \begin{bmatrix} (\mathbf{n}_i)_x \\ (\mathbf{n}_i)_y \\ (\mathbf{n}_i)_z \end{bmatrix}.$$

For the entire structure the F_i and Δ_i are assembled into the bar force matrix F and the bar displacement matrix Δ,

$$F = \begin{bmatrix} F_1 \\ F_2 \\ \vdots \\ F_B \end{bmatrix} \quad \text{and} \quad \Delta = \begin{bmatrix} \Delta_1 \\ \Delta_2 \\ \vdots \\ \Delta_B \end{bmatrix},$$

where B is the number of bars in the structure.

From elementary mechanics of solids (*see* Appendix A.3) it is known that the bar forces and displacements are related through Hooke's law,

$$F_i = K_i \Delta_i,$$

in which

$$K_i = \frac{A_i E}{L_i}. \tag{2.2}$$

(A_i is the area of the ith bar and E is Young's modulus.)

For the entire structure Eq. (2.2) becomes

$$K = \begin{bmatrix} K_1 & & & \\ & K_2 & & \bigcirc \\ & & \cdot & \\ & & & \cdot \\ \bigcirc & & & \cdot \\ & & & & K_B \end{bmatrix}. \tag{2.3}$$

The matrix K is called the primitive stiffness matrix.

The geometrical bar-displacement joint-displacement relationship, Eq. (2.1), can now be written for the entire structure through the aid of the matrix N, a $B \times J$ (row $\times$ column) matrix whose elements N_{ij} are

$$N_{ij} = \begin{cases} \tilde{n}_i & \text{if } j \text{ is the } + \text{ end of bar } i \\ -\tilde{n}_i & \text{if } j \text{ is the } - \text{ end of bar } i \\ 0 & \text{otherwise} \end{cases} \tag{2.4}$$

Using N, Eq. (2.1) becomes

$$\Delta = N\delta. \tag{2.5}$$

Note that here, as in the case of the matrix δ, it has been convenient to state the definitions in terms of *partitioned* matrices.

Comparing the definition of N with the definition of the matrix A of Chapter 1 provides the motivation for referring to N as a generalized branch node incidence matrix or simply an incidence matrix. Since N contains no columns corresponding to support joints, the implication is that all supports are "completely fixed" (i.e. if I is a support node, $\delta_I = 0$). It may also be noted that any partial support can be replaced by an equivalent fixed support as indicated in Fig. 2.5, but care should be taken not to ill-condition the system.

To complete the discussion of the node method for trusses, it remains only to discuss the joint equilibrium equations. It will now be shown that they can be written as

$$\tilde{N}F = P, \tag{2.6}$$

using matrices previously defined. Fig. 2.6 shows a typical truss joint whose equilibrium equation (the vector sum of all forces on any joint

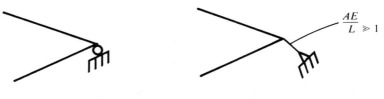

Partial support Equivalent fixed support

Fig. 2.5.

free-body diagram must be zero) is

$$F_\alpha \mathbf{n}_\alpha - F_\beta \mathbf{n}_\beta - F_j \mathbf{n}_j + F_\delta \mathbf{n}_\delta = -\mathbf{P}_i. \tag{2.7}$$

In general a joint equilibrium equation must contain a term $\pm F_I \mathbf{n}_I$ for each bar incident upon the joint — the sign is determined by whether the bar is positively or negatively incident. This is precisely the form provided by Eq. (2.6).

The node method for trusses then becomes

$$\tilde{N}F = P \qquad \text{(node equilibrium equation)}$$
$$F = K\Delta \qquad \text{(Hooke's law)}$$
$$\Delta = N\delta \qquad \text{(branch-displacement joint-displacement equation)}$$

from which it follows that

$$\tilde{N}F = P \quad \rightarrow \quad \tilde{N}K\Delta = P \quad \rightarrow \quad \tilde{N}KN\delta = P$$

or

$$\delta = (\tilde{N}KN)^{-1}P \tag{2.8}$$

Equation (2.8) is a system of simultaneous linear algebraic equations on the unknown joint displacements δ; once they have been computed it is a simple matter to compute Δ using Eq. (2.1), and F using Hooke's law.

The node method begins with forces which satisfy both equilibrium and Hooke's law and writes them in terms of the joint displacements. This insures that the three requirements for solutions are satisfied; in particular, the pieces fit together since their bar length changes are determined from joint displacements.

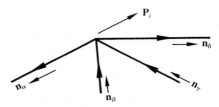

Fig. 2.6. Joint equilibrium.

Note that in the case of the equilibrium equations, the omission of support nodes from the matrix N results in not writing equilibrium equations at these nodes.

2.4 A DECOMPOSITION

At the heart of the node method is the matrix $\tilde{N}KN$, the *system matrix*. While the formal description of the node method for trusses is complete as it is given in Section 2.3, from a practical point of view it would be a mistake to program this formulation directly, largely because the matrices N and K are sparse. In this section, a method will be given by which the system matrix can be formed directly from the unit vectors $\mathbf{n}_i$, and the bar stiffnesses K_i without explicitly forming the matrices N and K.

To achieve this end it is convenient to partition the matrix N into its rows

$$N = \begin{bmatrix} N_1 \\ N_2 \\ \vdots \\ N_B \end{bmatrix}. \tag{2.9}$$

Since K is diagonal, it follows that the system matrix can be written as a sum,

$$\tilde{N}KN = \sum_{i=1}^{B} \tilde{N}_i K_i N_i. \tag{2.10}$$

The term $\tilde{N}_i K_i N_i$ can be regarded as the contribution of bar i to the system matrix.

Proceeding in this vein one more step, if neither end of bar i is a support, it follows from Eq. (2.4) (using the notation of Fig. 2.3) that N_i contains two non-zero terms and that $\tilde{N}_i K_i N_i$ contributes four terms

$$\tilde{N}_i K_i N_i = \begin{bmatrix} \vdots \\ n_i \\ \vdots \\ -n_i \\ \vdots \end{bmatrix} K_i [\cdots \tilde{n}_i \cdots -\tilde{n}_i \cdots]$$

$$= \begin{bmatrix} \overset{\text{col. } A}{\vert} & \overset{\text{col. } C}{\vert} \\ - n_i K_i \tilde{n}_i - & -n_i K_i \tilde{n}_i - \\ - -n_i K_i \tilde{n}_i - & n_i K_i \tilde{n}_i - \end{bmatrix} \begin{matrix} \text{row } A \\ \\ \text{row } C \end{matrix}$$

25

to the system matrix. When either end of bar i is a support node only one diagonal term is generated.

Figure 2.7 shows a simple 2-dimensional example which is used here to illustrate the decomposition of the system matrix into the contributions of each of the bars. It is assumed that the stiffness of each of the bars is known; the unit vectors are

$$\mathbf{n}_1 = \frac{1}{\sqrt{2}}\begin{bmatrix} 1 \\ 1 \end{bmatrix}, \qquad \mathbf{n}_2 = \frac{1}{\sqrt{2}}\begin{bmatrix} -1 \\ 1 \end{bmatrix}, \qquad \mathbf{n}_3 = \begin{bmatrix} 1 \\ 0 \end{bmatrix}, \qquad \mathbf{n}_4 = \begin{bmatrix} 0 \\ 1 \end{bmatrix}.$$

The matrix N is then

$$N = \begin{bmatrix} N_1 \\ N_2 \\ N_3 \\ N_4 \end{bmatrix} = \begin{bmatrix} \tilde{n}_1 & 0 \\ \tilde{n}_2 & -\tilde{n}_2 \\ 0 & \tilde{n}_3 \\ 0 & \tilde{n}_4 \end{bmatrix},$$

and the contribution of bar 2, e.g., to the system matrix is simply

$$\tilde{N}_2 K_2 N_2 = \begin{bmatrix} n_2 \\ -n_2 \end{bmatrix} K_2 [\tilde{n}_2 - \tilde{n}_2] = \begin{bmatrix} n_2 K_2 \tilde{n}_2 & -n_2 K_2 \tilde{n}_2 \\ -n_2 K_2 \tilde{n}_2 & n_2 K_2 \tilde{n}_2 \end{bmatrix}.$$

Since

$$n_2 K_2 \tilde{n}_2 = \frac{K_2}{2} \begin{bmatrix} 1 & -1 \\ -1 & 1 \end{bmatrix},$$

it follows that

$$\tilde{N}_2 K_2 N_2 = \frac{K_2}{2} \begin{bmatrix} \begin{bmatrix} 1 & -1 \\ -1 & 1 \end{bmatrix} & -\begin{bmatrix} 1 & -1 \\ -1 & 1 \end{bmatrix} \\ -\begin{bmatrix} 1 & -1 \\ -1 & 1 \end{bmatrix} & \begin{bmatrix} 1 & -1 \\ -1 & 1 \end{bmatrix} \end{bmatrix}.$$

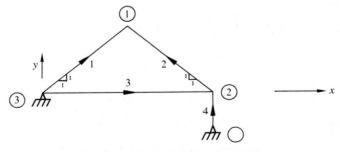

Fig. 2.7. A two dimensional example.

Adding the contributions of all four bars, the system matrix for the example is finally

$$\tilde{N}KN = \begin{bmatrix} \dfrac{K_1}{2}+\dfrac{K_2}{2} & \dfrac{K_1}{2}-\dfrac{K_2}{2} & \dfrac{K_2}{2} & \dfrac{K_2}{2} \\[2mm] \dfrac{K_1}{2}-\dfrac{K_2}{2} & \dfrac{K_1}{2}+\dfrac{K_2}{2} & \dfrac{K_2}{2} & \dfrac{K_2}{2} \\[2mm] \dfrac{K_2}{2} & \dfrac{K_2}{2} & \dfrac{K_2}{2}+K_3 & \dfrac{K_2}{2} \\[2mm] \dfrac{K_2}{2} & \dfrac{K_2}{2} & \dfrac{K_2}{2} & \dfrac{K_2}{2}+K_4 \end{bmatrix}$$

2.5 EXERCISES

Programs P.1 and P.2 included at the end of this book illustrate the generality of the work discussed in this chapter and provide additional numerical examples. Based upon these programs, the following exercises may be useful.

1. In order to test his understanding of the material presented thus far, the reader can attempt to reconstruct Program P.2, the space truss program starting from Program P.1, the plane truss program.
2. As an even more rigorous test of his understanding of the material, the reader can now anticipate Chapter 6 and modify Programs P.1 or P.2 to include thermal effects.
3. Modify Programs P.1 or P.2 to include "buckling" effects.

2.5.1 Discussion of Exercise 3

In linear structural analysis the equilibrium equations are written in the undeformed configuration. The classical treatment of overall truss buckling (not member buckling), on the other hand, adds to the linear formulation a nonlinearity which approximates the effect of small changes of geometry on the linear formulation in a manner similar to the method used to derive the linear response of a membrane or string. How this can be done is now indicated for a single bar one end of which is allowed to move. The general case is simply a composite in which each end of each bar is considered.

Figure 2.8 shows bar i in its undeformed and deformed configurations. Assuming that the bar force F_i remains approximately constant, the displacement δ_A generates a force normal to the undeformed position of the bar. Following the string problem this force has a magnitude of $F_i \times \theta$

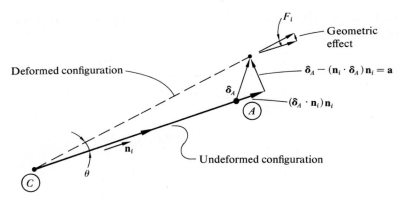

Fig. 2.8.

(*see* Fig. 2.8) of small θ. In vector form, at end A this force is

$$F_i \frac{\mathbf{a}}{|\mathbf{a}|} \frac{|\mathbf{a}|}{L_i} = \frac{F_i}{L_i}[\boldsymbol{\delta}_A - (\mathbf{n}_i \cdot \boldsymbol{\delta}_A)\mathbf{n}_i]$$

It only remains to introduce this term in an appropriate manner into Programs P.1 and P.2.

THREE

A General Statement of the Node Method

3.1 INTRODUCTION

In this chapter a rather general discussion of the node method for structures is presented. The intention is to provide a formalism under which any class of structures can be considered. The reader can judge for himself the extent to which this has been accomplished by observing the wide class of problems now treated using finite elements methods (*see* Azar or Zienkiewicz) and the high compatibility of these methods with the formalism presented here.

With the material of this chapter, the attempt is also to unify much of this text. While Chapters 2, 4, and 5 are in fact special applications of the material presented in this chapter, several of the developments in the other chapters depend heavily upon a clear understanding of the node method and therefore also motivate this chapter.

Before beginning the formal statement of the node method, it is perhaps in order to comment on the fact that such a large fraction of this book is devoted to it. In spite of the fact that experience indicates that the classical methods are easier to execute when working without a computer, some very large proportion of the computer programs in use today are based on the node method. The justification for this sacrifice in computational efficiency is simply that the node method is very easy to program. It is one of the delightful aspects of the state-of-the-art that it is now possible to achieve general computer programs with modest programming effort.

3.2 THE NODE METHOD

It is assumed that the structure to be analyzed has a well defined set of nodes and branches (or elements) and that two matrices are associated with the nodes,

P — joint load matrix,
δ — joint displacement matrix.

Associated with the branches (members) of the structure are

F — member force matrix,
Δ — member displacement matrix,
K — primitive stiffness matrix.

In terms of these quantities the equations of the node method are simply

$$\tilde{N}F = P \quad \text{(joint equilibrium)}$$
$$F = K\Delta \quad \text{(Hooke's law)} \tag{3.1}$$
$$\Delta = N\delta \quad \text{(member displacement-joint displacement equation)}$$

from which it follows that

$$\tilde{N}F = P \quad \rightarrow \quad \tilde{N}K\Delta = P \quad \rightarrow \quad \tilde{N}KN\delta = P$$

or

$$\delta = (\tilde{N}KN)^{-1}P \tag{3.2}$$

The analysis problem consists of finding, F, Δ, and δ given N, K, and P. Eq. (3.2) indicates that in the node method, δ is computed first from which Δ and F may be computed using the second and third parts of Eqs. (3.1). It should be noted that Eqs. (3.1 and 3.2) are the most simple statement of the node method; such effects as temperature, lack of fit, support settlement, etc., will be added in a later chapter.

Nothing in this formulation implies that the structure under discussion is composed of members with two ends. Quite to the contrary, this formulation is valid beyond its use in this text and can be applied to shell and solid finite elements.

Finally, some remarks are in order concerning the "incidence matrix", N. While Eqs. (3.1) simply describe a linear system, the description is peculiar in that it uses the matrix N twice. This multiple use of N is not necessary and is purely a matter of convenience which can be motivated most easily in terms of energy.

Let the work done by the external loads and the internal strain energy be defined as

$$W = \tfrac{1}{2}\tilde{P}\delta \quad \text{and} \quad E = \tfrac{1}{2}\tilde{F}\Delta, \tag{3.3}$$

respectively. If the node method is to describe a "conservative system" then

$$W = E \quad \text{or} \quad \tilde{P}\delta = \tilde{F}\Delta. \tag{3.4}$$

The following theorem can now be proved.

THEOREM: *If $W = E$ and $\tilde{N}F = P$ are to hold for arbitrary F, it follows that $\Delta = N\delta$.*

Proof: Since $\tilde{N}F = P$, Eq. (3.4) can be written as

$$\tilde{F}N\delta = \tilde{F}\Delta. \tag{3.5}$$

If Eq. (3.5) is to be valid for an arbitrary F, then $\Delta = N\delta$. If it were not, there would exist a component of Δ, say Δ_I for which

$$(N\delta)_I \neq \Delta_I,$$

which would yield a contradiction to Eq. (3.5) by selecting the elements of F to be zero except $F_I = 1$.

If some form other than Eq. (3.1) is used for the node method, it is simply necessary to use some form other than Eq. (3.3) for the energy.

3.3 THE DECOMPOSITION

With numerical computation in mind, it is worthwhile to pursue the details of the system matrix, $\tilde{N}KN$, a little further. The most striking feature of the matrices N and K is that they are sparse; this sparseness discourages their direct formation in the computer. In this section a scheme is discussed which allows the formation of $\tilde{N}KN$ from the elements of N and K without explicitly constructing the matrices themselves. It is here that use is made of the fact that each member has two ends.

The first step in the decomposition is to partition the matrix N into rows, each of which corresponds to a member. (In general these rows themselves are partitioned matrices and not single rows of elements as they were in the case of the truss.) This partitioning of N implies a partitioning of K. Since the elements (branches) of the structure are independent, the matrix K is "partitioned diagonal". N and K therefore have the form

$$N = \begin{bmatrix} N_1 \\ N_2 \\ \vdots \\ N_B \end{bmatrix}, \quad K = \begin{bmatrix} K_1 & & & \bigcirc \\ & K_2 & & \\ & & \cdot & \\ & & & \cdot & \\ \bigcirc & & & & K_B \end{bmatrix},$$

in which B is the number of branches (elements) in the structure. Since K is diagonal, the *system matrix* $\tilde{N}KN$ becomes a sum,

$$\tilde{N}KN = \sum_{i=1}^{b} \tilde{N}_i K_i N_i,$$

in which the term $\tilde{N}_i K_i N_i$ is the contribution of member i to the system matrix. Looking a little more closely at the structure of N and using the fact that each branch has two ends, for a member neither end of which is a support N_i has the form

$$
\begin{array}{cc}
\text{col. } A & \text{col. } B \\
\end{array}
$$
$$
N_i = [\cdots \eta_i^+ \cdots \eta_i^- \cdots]
$$

and

$$
\tilde{N}_i K_i N_i =
\begin{array}{cc}
\text{col. } A & \text{col. } B \\
\end{array}
\left[
\begin{array}{cc}
-\tilde{\eta}_i^+ K_i \eta_i^+ - & \tilde{\eta}_i^+ K_i \eta_i^- - \\
-\tilde{\eta}_i^- K_i \eta_i^+ - & \tilde{\eta}_i^- K_i \eta_i^- -
\end{array}
\right]
\begin{array}{c}
\text{row } A \\
\text{row } B
\end{array}
$$

in which A and B are the positive and negative ends of member i respectively.

It follows that, in general, each member contributes four terms to the system matrix. These terms can be placed directly into the system matrix and it is not necessary to construct N and K explicitly. The computer programs at the end of this text illustrate this procedure.

3.4 EXERCISES

1. Show how the decomposition of Section 3.3 generalizes for members with three and four nodes (triangular and quadrilateral finite elements).
2. Show how the formulation of the node method changes when the order of the components of either, but not both, of the node force matrix or the node displacement matrix is changed.

   Answer: Since the truss is the only structure already discussed in detail in this text, it will be used to discuss this exercise. A more meaningful discussion would involve the selection of, e.g., member displacements for the frame.

   If, for the truss, the joint displacement matrix δ were defined as it is in the preceding chapter but the joint force matrix were defined to have other components, say

$$
\delta_i =
\begin{bmatrix}
(\delta_i)_x \\
(\delta_i)_y \\
(\delta_i)_z
\end{bmatrix}
\quad \text{and} \quad
P_i =
\begin{bmatrix}
(P_i)_z \\
(P_i)_y \\
(P_i)_x
\end{bmatrix},
$$

it would be necessary to use a permutation matrix to maintain the physical interpretation of work and energy. This might be done using the matrix A,

$$A = \begin{bmatrix} A_1 & & & \\ & A_2 & & \bigcirc \\ & & \cdot & \\ & & & \cdot \\ \bigcirc & & & \cdot \\ & & & A_J \end{bmatrix},$$

in which

$$A_i = \begin{bmatrix} & + & 1 \\ + & 1 & + \\ 1 & + & \end{bmatrix} \quad \text{(for any } i\text{)}$$

has been selected so that

$$\mathbf{P}_i \cdot \boldsymbol{\delta}_i = (\mathbf{P}_i)_x (\boldsymbol{\delta}_i)_x + (\mathbf{P}_i)_y (\boldsymbol{\delta}_i)_y + (\mathbf{P}_i)_z (\boldsymbol{\delta}_i)_z = \tilde{\delta}_i A_i P_i,$$

or

$$W = \tfrac{1}{2} \tilde{P} \tilde{A} \delta.$$

The node method then becomes

$\tilde{N}F = AP$ (equilibrium)

$F = K\Delta$ (Hooke's law)

$\Delta = N\delta$ (member displacement-joint displacement)

But since $A^{-1} = \tilde{A}$, the equilibrium equation can be modified so that the equations of the node method read,

$\tilde{B}F = P$

$F = K\Delta$

$\Delta = N\delta$

in which

$B = NA,$

a formulation in which repeated use is not made of the matrix N.

FOUR

The Node Method for
Plane Frames

4.1 THE PLANE FRAME

In order of increasing complexity, the plane frame provides a convenient step up from the truss. Here the concepts of local and global coordinates, which were hidden in the truss problem, become explicit while the number of quantities required for the description of a beam element (3) is sufficiently small to allow simple figures to be drawn.

The plane frame considered in this chapter is a skeletal structure constructed by assembling plane beams using rigid connections. It is loaded only at its nodes and does not undergo temperature effects, lack of fit, etc.; like the truss it has completely "fixed" supports (i.e. at a support both the displacement vector and the rotation are assumed to be zero). Briefly, it will be shown later that the assumption of rigid connections which is not always realized in actual construction is in fact not a restriction; loads acting along members, temperature effects, lack of fit, etc., are discussed in a later chapter and are shown to be secondary from a computational point of view. The use of fixed supports is consistent with the use of rigid connection and is, again, not a restriction on the formulation (*see* Fig. 4.1). Here, as in the other chapters, the payoff for a careful problem statement is a simple formulation.

The rigid connection of elements is the key to the difference between plane frames and plane trusses. It allows adjacent members to restrain each others' end rotations giving rise to moments at the ends of members which do not exist in trusses. In the same vein, associated with each (rigid) joint in a plane frame is a single scalar which describes its rotation; no such quantity exists for the plane truss.

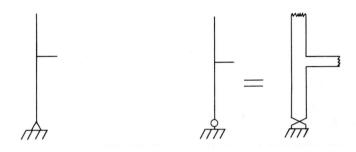

Frame with hinged support   Equivalent fixed support
                                   with member release

Fig. 4.1. Releases.

4.2 ROTATION OF COORDINATES

This chapter and subsequent chapters make use of the well known transformation of vector components under rotation of coordinates. While the case of three dimensions is outlined in Appendix 1, plane vectors are particularly simple and are treated in this section geometrically.

Figure 4.2 shows a typical plane vector **t** which may be regarded as a position vector. It also shows two coordinate systems, the primed and the unprimed, which differ in orientation by the angle ϕ. Let

$(\mathbf{t})_x$ — x component of the vector **t** in the unprimed coordinate system
$(\mathbf{t})_y$ — y component of the vector **t** in the unprimed coordinate system
$(\mathbf{t})'_x$ — x component of vector **t** in the primed coordinate system
$(\mathbf{t})'_y$ — y component of vector **t** in the primed coordinate system

With the aid of the segment $(\mathbf{t})_x \sin \phi$ shown in Fig. 4.2 and other segments like it, the geometrical relationships between the components of **t** in the two systems follow as:

$$(\mathbf{t})'_x = (\mathbf{t})_x \cos \phi + (\mathbf{t})_y \sin \phi$$
$$(\mathbf{t})'_y = -(\mathbf{t})_x \sin \phi + (\mathbf{t})_y \cos \phi$$

which are written in matrix notation in Appendix 1 as

$$t' = At \quad \text{or} \quad t = \tilde{A}t',$$

using the fact that rotation matrices are orthogonal, i.e. $A^{-1} = \tilde{A}$. Based upon this relationship, a hybrid rotation matrix (which contains a third component which does not transform) is constructed in the next section; Chapter 5 uses the full three dimensional form indicated in the appendix.

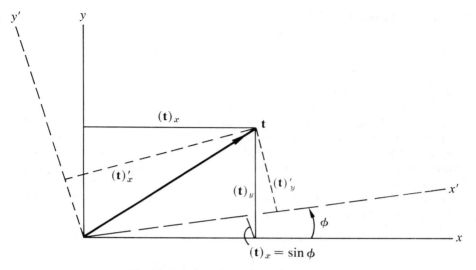

Fig. 4.2. Rotation of coordinates in the plane.

4.3   A FORMAL DESCRIPTION OF THE PLANE FRAME

Figure 4.3 shows a typical joint of a plane frame. Since it is assumed that the members are rigidly connected, all member ends which meet at a joint have the same x and y displacement components and the same rotation. Proceeding more formally, associated with joint i are an applied force vector $\mathbf{P}_i$, applied moment M_i, a displacement vector $\boldsymbol{\delta}_i$ and a rotation θ_i. These are again described for the entire structure by the matrices

$$\boldsymbol{\delta} = \begin{bmatrix} \delta_1 \\ \delta_2 \\ \vdots \\ \delta_J \end{bmatrix} \quad \text{and} \quad P = \begin{bmatrix} P_1 \\ P_2 \\ \vdots \\ P_J \end{bmatrix},$$

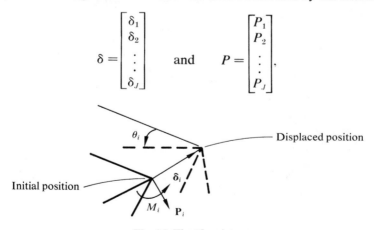

Fig. 4.3. The ith point.

37

but here the elements are

$$\delta_i = \begin{bmatrix} (\delta_i)_X \\ (\delta_i)_Y \\ \theta_i \end{bmatrix} \quad \text{and} \quad P_i = \begin{bmatrix} (\mathbf{P}_i)_X \\ (\mathbf{P}_i)_Y \\ M_i \end{bmatrix},$$

J again is the number of movable joints.

The member description for the plane frame is considerably more complicated than that of the truss, as can be seen in Fig. 4.4. In the plane, a beam is acted upon by two force components and one moment component at each end, a total of six quantities. But since these six quantities are related by three equations of equilibrium, there are three quantities which may be specified arbitrarily. By this argument it is determined that the member force matrix will contain three components for each member in the structure, or

$$F = \begin{bmatrix} F_1 \\ F_2 \\ \cdot \\ \vdots \\ F_B \end{bmatrix} \quad \text{and} \quad F_i = \begin{bmatrix} t_i \\ m_i^+ \\ m_i^- \end{bmatrix}.$$

It should be noted that there is a certain arbitrariness in the selection of the elements of F_i. Actually any three of the force or moment components

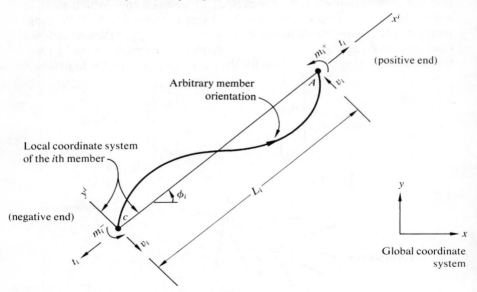

Fig. 4.4. The ith member.

shown in Fig. 4.4 may be used, provided that it is possible to compute the other three from them using the equations of equilibrium. The above choice of F_i is motivated at least partially by the fact that it allows the direct use of stiffness coefficients developed for members with a variable moment of inertia for use with the moment distribution method.

Figure 4.4 also provides the first introduction to the concepts of local and global coordinates. Since the stiffness of a member is a function of its orientation in space, in a global coordinate system (i.e. a coordinate system fixed in space) identical members which are oriented differently have different stiffnesses. In order to isolate the effect of member orientation on member stiffness, or said differently, in order that identical members may have identical "primitive" stiffnesses, the concept of a local coordinate system is introduced. In its local coordinate system, each member lies along the x axis; it is therefore necessary to perform a rotation of coordinates to obtain the member properties in the global coordinates in which the equilibrium equations are written. Let

$$f_i^+ = \begin{bmatrix} t_i \\ v_i \\ m_i^+ \end{bmatrix} \quad \text{and} \quad f_i^- = \begin{bmatrix} -t_i \\ -v_i \\ m_i^- \end{bmatrix},$$

represent the end of member forces at the positive and negative ends of member i respectively. Using equilibrium, they can be written in terms of the member force, F_i, as

$$f_i^+ = N_i^+ F_i \quad \text{and} \quad f_i^- = \tilde{N}_i^- F_i \tag{4.1}$$

for which

$$N_i^+ = \begin{bmatrix} 1 & 0 & 0 \\ 0 & -L_i^{-1} & 1 \\ 0 & -L_i^{-1} & 0 \end{bmatrix} \quad \text{and} \quad N_i^- = \begin{bmatrix} -1 & 0 & 0 \\ 0 & L_i^{-1} & 0 \\ 0 & L_i^{-1} & 1 \end{bmatrix}.$$

The end of member forces in the global coordinate system are now obtained by multiplying f_i^+ and f_i^- on the left by the transpose of the rotation matrix,

$$R_i = \begin{bmatrix} \cos \phi_i & \sin \phi_i & 0 \\ -\sin \phi_i & \cos \phi_i & 0 \\ 0 & 0 & 1 \end{bmatrix}.$$

Since the member quantities have in this section been described thus far in terms of forces, it is consistent to use the equilibrium equations, $\tilde{N}F = P$, to define the matrix N.

For a plane frame there are three scalar equilibrium equations associated with each joint. The vector sum of all forces on any joint free body diagram must be zero (two scalar equations), and the sum of the moments on any joint free body diagram must be zero (one scalar equation). In terms of the notation just developed, a joint matrix equilibrium equation must contain either $\tilde{R}_i\tilde{N}_i^+F_i$ or the term $\tilde{R}_i\tilde{N}_i^-F_i$ for each bar i incident upon the joint—the former when the bar i is positively incident upon the joint under consideration and the latter when it is negatively incident. It follows that the matrix N is a $B \times J$ (row × column) matrix whose elements are

$$N_{ij} = \begin{cases} N_i^+R_i & \text{if } j \text{ is the positive end of member } i \\ N_i^-R_i & \text{if } j \text{ is the negative end of member } i \\ 0 & \text{otherwise} \end{cases}$$

While in the case of the truss there is very little freedom of choice available in the selection of the member displacement Δ_i, for the plane frame, it is not immediately obvious what physical quantities should constitute Δ_i. However, the formulation used takes care of that automatically. Since the matrices N and δ have already been defined for the plane frame, the equation $\Delta = N\delta$ can be used to interpret Δ. Writing

$$\Delta = \begin{bmatrix} \Delta_1 \\ \Delta_2 \\ \vdots \\ \Delta_B \end{bmatrix} \qquad \text{where} \qquad \Delta_i = \begin{bmatrix} \Delta L_i \\ \alpha_i^+ \\ \alpha_i^- \end{bmatrix},$$

it is easily shown that

$$\Delta_i = N_i^+R_i\delta_A + N_i^-R_i\delta_C. \tag{4.2}$$

Note first that in Eq. (4.2) the terms $R_i\delta_A$ and $R_i\delta_C$ are simply the values of the joint displacements at joints A and C viewed in the local coordinate system associated with member i. Let the components in the local coordinate system be indicated by the subscripts x^i and y^i. It is now possible to perform the multiplications indicated by Eq. (4.2) which gives

$$\Delta_i = \begin{bmatrix} \Delta L_i \\ \alpha_i^+ \\ \alpha_i^- \end{bmatrix} = \begin{bmatrix} (\delta_A)_{x^i} - (\delta_C)_{x^i} \\ \theta_A - \dfrac{1}{L_i}[(\delta_A)_{y^i} - (\delta_C)_{y^i}] \\ \theta_C - \dfrac{1}{L_i}[(\delta_A)_{y^i} - (\delta_C)_{y^i}] \end{bmatrix},$$

from which it follows that

ΔL_i — the length change of the ith member
α_i^+ — the rotation of the positive end less the rigid body rotation of the member
α_i^- — the rotation of the negative end less the rigid body rotation of the member

It remains to discuss the primitive stiffness matrix for plane frames. Since this is discussed quite generally in Chapter 7, only the case of uniform straight members will be discussed here. Probably the most direct way of viewing the stiffness matrix comes through the introduction of unit values of the elements of the member displacement matrix Δ_i for which the associated forces become elements of K_i. Figure 4.5 indicates a simple example of this procedure in which

$$\Delta_i = \begin{bmatrix} 0 \\ 1 \\ 0 \end{bmatrix} \quad \text{for which} \quad F_i = K_i \Delta_i = \begin{bmatrix} (K_i)_{12} \\ (K_i)_{22} \\ (K_i)_{32} \end{bmatrix}.$$

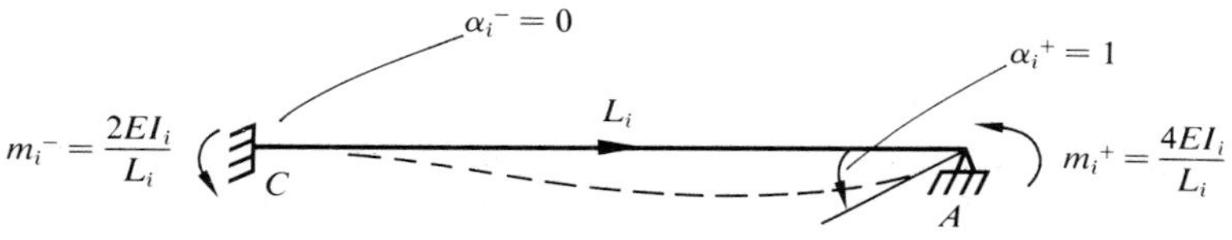

$$\alpha_i^- = 0$$

$$\alpha_i^+ = 1$$

$$m_i^- = \frac{2EI_i}{L_i}$$

$$L_i$$

$$m_i^+ = \frac{4EI_i}{L_i}$$

C A

Fig. 4.5. The stiffness matrix.

For a uniform straight beam in bending the values of F_i are shown in Fig. 4.3 and derived in Appendix A.3. The stiffness matrix K is then simply

$$K = \begin{bmatrix} K_1 & & & \\ & K_2 & & \bigcirc \\ & & \ddots & \\ & \bigcirc & & K_B \end{bmatrix} \quad \text{where} \quad K_i = \begin{bmatrix} \dfrac{A_i}{L_i} & & \\ & \dfrac{4I_i}{L_i} & \dfrac{2I_i}{L_i} \\ & \dfrac{2I_i}{L_i} & \dfrac{4I_i}{L_i} \end{bmatrix} E$$

and

A_i — area of the ith member
L_i — length of the ith member
I_i — moment of inertia of the ith member

4.4 A DECOMPOSITION

Following Section 4.3, it is again possible to identify the contribution of each member to the system matrix. Let

$$N = \begin{bmatrix} N_1 \\ N_2 \\ \vdots \\ N_B \end{bmatrix} \quad \text{and} \quad \tilde{N}KN = \sum_{i=1}^{B} \tilde{N}_i K_i N_i,$$

the contribution of the ith member to the system matrix, $\tilde{N}KN$, is

$$\tilde{N}_i K_i N_i = \begin{bmatrix} \vdots \\ \tilde{R}_i \tilde{N}_i^+ \\ \vdots \\ \tilde{R}_i \tilde{N}_i^- \end{bmatrix} K_i [\cdots N_i^+ R_i \cdots N_i^- R_i \cdots]$$

$$= \begin{bmatrix} \overset{\text{col } A}{\big|} & \overset{\text{col. } C}{\big|} \\ -\tilde{R}_i \tilde{N}_i^+ K_i N_i^+ R_i - & \tilde{R}_i \tilde{N}_i^+ K_i N_i^- R_i - \\ \big| & \big| \\ -\tilde{R}_i \tilde{N}_i^- K_i N_i^+ R_i - & \tilde{R}_i \tilde{N}_i^- K_i N_i^- R_i - \\ \big| & \big| \end{bmatrix} \begin{matrix} \text{row } A \\ \\ \text{row } C \end{matrix}$$

Program P.3 included at the end of this book illustrates the work discussed in this chapter and provides a numerical example.

4.5 A SIMPLE EXAMPLE

Figure 4.6 shows an example for which the matrices just presented will be described in detail. For simplicity let the member stiffnesses be

$$K_1 = K_2 = K_3 = \begin{bmatrix} 1 & 0 & 0 \\ 0 & 1 & \frac{1}{2} \\ 0 & \frac{1}{2} & 1 \end{bmatrix}$$

Furthermore, let $L_1 = L_2 = L_3 = 1$. Since $\phi_1 = \phi_3 = 0$, $R_1 = R_3 = I$ and

$$R_2 = \begin{bmatrix} 1/\sqrt{2} & 1/\sqrt{2} & 0 \\ -1/\sqrt{2} & 1/\sqrt{2} & 0 \\ 0 & 0 & 1 \end{bmatrix}$$

Following the decomposition described in Section 4.4, the system matrix can be written directly in partitioned form as

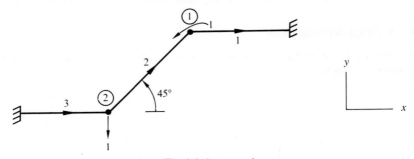

Fig. 4.6. An example.

$$
\tilde{N}KN = \begin{bmatrix} \tilde{R}_2\tilde{N}_2{}^+K_2N_2{}^+R_2 & \tilde{R}_2\tilde{N}_2{}^+K_2N_2{}^-R_2 \\ + & \\ \tilde{R}_1\tilde{N}_1{}^-K_1N_1{}^-R_1 & \\ \hline \tilde{R}_2\tilde{N}_2{}^-K_2N_2{}^+R_2 & \tilde{R}_2\tilde{N}_2{}^-K_2N_2{}^-R_2 \\ & + \\ & \tilde{R}_3\tilde{N}_3{}^+K_3N_3{}^+R_3 \end{bmatrix}
$$

Which after multiplication is

$$
\tilde{N}KN = \begin{bmatrix}
2+1 & -1 & \dfrac{3}{2\sqrt{2}} & -1 & 0 & 0 \\[2ex]
-1 & 2+3 & -\dfrac{3}{2\sqrt{2}}+\dfrac{3}{2} & 0 & -3 & -\tfrac{3}{2} \\[2ex]
\dfrac{3}{2\sqrt{2}} & -\dfrac{3}{2\sqrt{2}}+\dfrac{3}{2} & 1+1 & 0 & \tfrac{3}{2} & \tfrac{1}{2} \\[2ex]
-1 & 0 & 0 & 2+1 & -1 & -\dfrac{3}{2\sqrt{2}} \\[2ex]
0 & -3 & \tfrac{3}{2} & -1 & 2+3 & \dfrac{3}{2\sqrt{2}}-\dfrac{3}{2} \\[2ex]
0 & -\tfrac{3}{2} & \tfrac{1}{2} & -\dfrac{3}{2\sqrt{2}} & \dfrac{3}{2\sqrt{2}}-\dfrac{3}{2} & 1+1
\end{bmatrix}
$$

The joint load and joint displacement matrices are

$$
P = \begin{bmatrix} (\mathbf{P}_1)_x \\ (\mathbf{P}_1)_y \\ \\ (\mathbf{P}_2)_x \\ (\mathbf{P}_2)_y \\ M_2 \end{bmatrix} = \begin{bmatrix} 0 \\ 0 \\ 1 \\ 0 \\ -1 \\ 0 \end{bmatrix}, \quad \delta = \begin{bmatrix} (\boldsymbol{\delta}_1)_x \\ (\boldsymbol{\delta}_1)_y \\ \theta_1 \\ (\boldsymbol{\delta}_2)_x \\ (\boldsymbol{\delta}_2)_y \\ \theta_2 \end{bmatrix}
$$

4.6 EXERCISES

1. Using the notation of the node method, construct a formulation for "plane grids" (i.e. structures whose elements not only lie in a plane but also have negligible length changes).

2. Anticipating Chapter 7, modify Program P.3 to make it capable of treating members with variable moment of inertia by reading in the required stiffness coefficients (e.g. from *Handbook of Frame Constants*, Portland Cement Assoc., Chicago).

3. Indicate the modifications required in the formulation when an end of member force is used as the member force (e.g. when $F_i = f_i^+$).

4. Modify computer Program P.3 to perform plastic analysis.

Answer: In this exercise, the simplest plastic analysis problem is considered in which:

1. Loads are only allowed to act at nodes.
2. All supports are fixed.
3. The reduction of the moment capacity of a beam due to the presence of axial load is neglected.
4. The structure is subjected to proportional loading, i.e. $P = \lambda P_0$, where P is the usual joint load matrix, P_0 is a matrix which represents the fixed ratios of the loads, and $\lambda > 0$ is a scalar.
5. The members are uniform between nodes.

For this problem (*see* P. Hodge, *Plastic Analysis of Structures*, McGraw-Hill, 1959), one possible formulation is through *linear programming*:

Maximize λ
Subject to:

$$\tilde{N}F = \lambda P_0 \qquad \text{(equilibrium)}$$
$$|m_i^+| \leq \mu_i \qquad \text{(the bending moment}$$
$$|m_i^-| \leq \mu_i \qquad \text{diagram must be safe)}$$

where

μ_i — capacity of member i

In this formulation the collapse load is the largest for which it is possible to satisfy equilibrium and not exceed the capacity of any member.

The details of the solution of this problem are contained in computer Program P.8. This exercise indicates that once the capability of writing the equilibrium equations for an arbitrary structure is achieved, also achieved are concomitant capabilities which go beyond simple elastic analysis.

FIVE

The Node Method for Space Frames

5.1 THE SPACE FRAME

The space frame constitutes the final step of increasing complexity in class of structures made in this book. It introduces no new concepts over the plane frame, only complications. For that reason this chapter might well be omitted from an elementary undergraduate course.

A space frame is defined here to be a skeletal structure constructed by joining with rigid connections elements which are arbitrary curved beams. Again, the rigid connection of elements will be seen to imply no restrictions upon the generality of the formulation, and only loads which are applied to joints will be considered.

Briefly, the space frame differs from the plane frame only in dimension:

1. The displacement vector associated with each joint has three rather than two components.

2. The rotation vector associated with each joint has three components rather than one.

3. There are, therefore, six equilibrium equations associated with each joint rather than three.

4. The member displacement matrix has six rather than three components.

Note finally that corresponding to each of the above statements concerned with displacements is a statement concerned with forces and that these force and displacement comments have rather obvious ramifications with regard to the matrices N and K.

Since no new concepts are developed, this chapter moves along rather quickly and is not easily taken without the background developed in the preceding chapters.

5.2 A FORMAL DESCRIPTION OF THE SPACE FRAME

Figure 5.1 shows a typical frame joint associated with which are an applied force vector $\mathbf{P}_i$, an applied moment vector $\mathbf{m}_i$, a displacement vector $\boldsymbol{\delta}_i$, and a rotation vector $\boldsymbol{\theta}_i$. These are again described for the entire structure by the matrices

$$P = \begin{bmatrix} P_1 \\ P_2 \\ \vdots \\ P_J \end{bmatrix} \quad \text{and} \quad \delta = \begin{bmatrix} \delta_1 \\ \delta_2 \\ \vdots \\ \delta_J \end{bmatrix},$$

but here the elements are

$$\delta_i = \begin{bmatrix} (\boldsymbol{\delta}_i)_x \\ (\boldsymbol{\delta}_i)_y \\ (\boldsymbol{\delta}_i)_z \\ (\boldsymbol{\theta}_i)_x \\ (\boldsymbol{\theta}_i)_y \\ (\boldsymbol{\theta}_i)_z \end{bmatrix} \quad \text{and} \quad P_i = \begin{bmatrix} (\mathbf{P}_i)_x \\ (\mathbf{P}_i)_y \\ (\mathbf{P}_i)_z \\ (\mathbf{m}_i)_x \\ (\mathbf{m}_i)_y \\ (\mathbf{m}_i)_z \end{bmatrix};$$

J again is the number of movable joints.

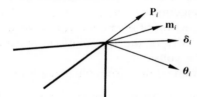

Fig. 5.1. The ith point.

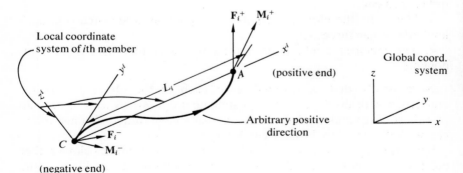

Fig. 5.2. The ith member.

The member description of a beam in space is considerably more complicated than that of a plane beam and for that reason it is shown rather schematically in Fig. 5.2. In its own local coordinate system, each member is again oriented along the x axis.

In space, a beam is acted upon by a force vector and a moment vector at each end, a total of twelve quantities. But since these twelve quantities are related by the six equilibrium equations of a rigid body in space, there are six quantities which may be specified arbitrarily. By this argument it is determined that the member force matrix will contain six components for each member in the structure, or

$$F = \begin{bmatrix} F_1 \\ F_2 \\ \vdots \\ \vdots \\ F_B \end{bmatrix} \quad \text{and} \quad F_i = \begin{bmatrix} (\mathbf{F}_i^+)_{x^i} \\ (\mathbf{M}_i^+)_{x^i} \\ (\mathbf{M}_i^+)_{y^i} \\ (\mathbf{M}_i^+)_{z^i} \\ (\mathbf{M}_i^-)_{y^i} \\ (\mathbf{M}_i^-)_{z^i} \end{bmatrix}$$

in which the subscripts x^i, y^i, z^i indicate components in the local coordinate system of the ith member. There is an arbitrariness in the selection of the components of F_i here, just as there is in the case of the plane frame. (The reader is referred to the comments given in Section 4.3.) It may be noted that the components of F_i can be classified roughly as a thrust, a twisting moment, and four bending moments.

Let

$$f_i^+ = \begin{bmatrix} (\mathbf{F}_i^+)_{x^i} \\ (\mathbf{F}_i^+)_{y^i} \\ (\mathbf{F}_i^+)_{z^i} \\ (\mathbf{M}_i^+)_{x^i} \\ (\mathbf{M}_i^+)_{y^i} \\ (\mathbf{M}_i^+)_{z^i} \end{bmatrix} \quad \text{and} \quad f_i^- = \begin{bmatrix} (\mathbf{F}_i^-)_{x^i} \\ (\mathbf{F}_i^-)_{y^i} \\ (\mathbf{F}_i^-)_{z^i} \\ (\mathbf{M}_i^-)_{x^i} \\ (\mathbf{M}_i^-)_{y^i} \\ (\mathbf{M}_i^-)_{z^i} \end{bmatrix}$$

represent the end of member forces at the positive and negative ends of member i respectively. Using equilibrium they can be written in terms of the member force, F_i, as

$$f_i^+ = \tilde{N}_i^+ F_i \quad \text{and} \quad f_i^- = \tilde{N}_i^- F_i \tag{5.1}$$

for which

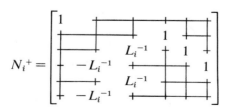

$$N_i^+ = \begin{bmatrix} 1 & & & & & \\ & & & & 1 & \\ & & L_i^{-1} & & & 1 \\ & -L_i^{-1} & & & & 1 \\ & & L_i^{-1} & & & \\ & -L_i^{-1} & & & & \end{bmatrix}$$

and

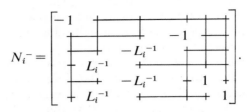

The end of member forces in the global coordinate system are now obtained by multiplying f_i^+ and f_i^- on the left by the transpose of the rotation matrix

$$R_i = \begin{bmatrix} A_i & 0 \\ 0 & A_i \end{bmatrix}.$$

in which A_i is the rotation matrix (*see* Appendix 1) which transforms a vector in the global coordinate system into a vector in the local co-ordinate system of the ith member; e.g.

$$(F_i^+)_{x^i} = A_{11}(F_i^+)_x + A_{12}(F_i^+)_y + A_{13}(F_i^+)_z.$$

Using the equilibrium equations. $NF = P$, again to define N, it follows that N is a $B \times J$ (row $\times$ column) matrix whose elements are

$$N_{ij} = \begin{cases} N_i^+ R_i & \text{if } j \text{ is the positive end of member } i \\ N_i^- R_i & \text{if } j \text{ is the negative end of member } i \\ 0 & \text{otherwise} \end{cases}$$

Since $\Delta = N\delta$, the member displacement matrix Δ is now defined and it only remains to interpret its components. From direct calculation it follows that

$$\Delta_i = \begin{bmatrix} (\delta_A)_{x^i} - (\delta_c)_{x^i} \\ (\theta_A)_{x^i} - (\theta_c)_{x^i} \\ (\theta_A)_{y^i} + [(\delta_A)_{z^i} - (\delta_c)_{z^i}]L_i^{-1} \\ (\theta_A)_{z^i} - [(\delta_A)_{y^i} - (\delta_c)_{y^i}]L_i^{-1} \\ (\theta_c)_{y^i} + [(\delta_A)_{z^i} - (\delta_c)_{z^i}]L_i^{-1} \\ (\theta_c)_{z^i} - [(\delta_A)_{y^i} - (\delta_c)_{y^i}]L_i^{-1} \end{bmatrix}$$

where

$$\Delta = \begin{bmatrix} \Delta_1 \\ \Delta_2 \\ \vdots \\ \Delta_B \end{bmatrix}$$

or simply

$$\Delta_i = N_i^+ R_i \delta_A + N_i^- R_i \delta_c \tag{5.2}$$

Finally, while the stiffness matrix K for arbitrary curved members is discussed in a later chapter, for uniform straight members it is simply

$$K = \begin{bmatrix} K_1 & & & & \\ & K_2 & & \bigcirc & \\ & & \cdot & & \\ & & & \cdot & \\ & \bigcirc & & & \cdot \\ & & & & K_B \end{bmatrix}$$

in which

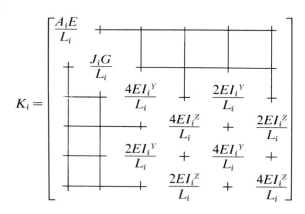

$$K_i = \begin{bmatrix} \dfrac{A_iE}{L_i} & & & & & \\ & \dfrac{J_iG}{L_i} & & & & \\ & & \dfrac{4EI_i^Y}{L_i} & & \dfrac{2EI_i^Y}{L_i} & \\ & & & \dfrac{4EI_i^Z}{L_i} & & \dfrac{2EI_i^Z}{L_i} \\ & & \dfrac{2EI_i^Y}{L_i} & & \dfrac{4EI_i^Y}{L_i} & \\ & & & \dfrac{2EI_i^Z}{L_i} & & \dfrac{4EI_i^Z}{L_i} \end{bmatrix}$$

and

L_i — length of ith member (projection on the x axis)
A_i — area of ith member
J_i — torsional stiffness of the ith member
I_i^Y — bending stiffness about y axis of the ith member
I_i^Z — bending stiffness about z axis of the ith member
E — Young's modulus
G — modulus of rigidity

5.3 A DECOMPOSITION

The system matrix may be decomposed into the sum of the contributions of the individual members precisely as was done in Section 4.4 for plane frames.

A computer program illustrating the remarks of this chapter is included at the end of the book.

SIX

Some Generalizations

6.1 INTRODUCTION

In this chapter, several complications which were omitted from the preceding chapters in the interest of simplicity are added to the formulation. It will be seen that these complications at the very worst require an additional step—but a step which is definitely secondary from a computational point of view.

6.2 MEMBER LOADS, TEMPERATURE, LACK OF FIT, etc.

With the exception of the truss, structures generally have loads applied at points other than nodes as indicated in Fig. 6.1. However, it will now be shown that under a simple decomposition, problems with member loads can be reduced to a primary problem which contains only node loads and a simple secondary problem. The secondary problem is obtained by first fixing all joints (setting $\delta = 0$); but in order to fix the joints, it is necessary to introduce fictitious joint forces, $\bar{P}$. (Note that the computation of $\bar{P}$ is a "local" problem not requiring the solution of a large system of simultaneous equations.) To this secondary solution which gives you the fictitious loads $\bar{P}$ must be added a solution to the problem in which loads $-\bar{P}$ are applied to the joints and the member loads are absent. This is the primary problem. This procedure is shown schematically in Fig. 6.2 for a plane frame with members loads.

Temperature, lack of fit, settlement of supports, etc., can be handled in a similar fashion as indicated for the case of a truss subjected to temperature change in Fig. 6.3.

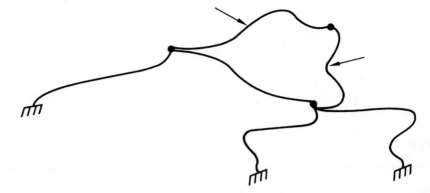

Fig. 6.1. Member loads.

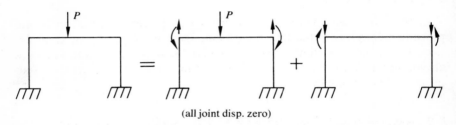

(all joint disp. zero)

Secondary problem   Primary problem

Fig. 6.2. Member loads.

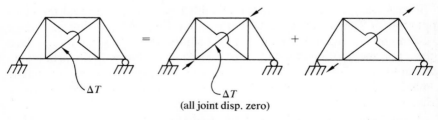

(all joint disp. zero)

Secondary problem   Primary problem

Fig. 6.3. Temperature effects in a truss.

6.3 EXAMPLES

Figure 6.4 indicates a very simple member load problem which, while not quite in the spirit of this book, (it includes a support which is not completely fixed), illustrates the decomposition nicely. EI is assumed to be constant.

Figure 6.5 shows a temperature problem to which the decomposition has been applied. In Fig. 6.6 a support problem is treated indirectly as a temperature problem.

6.4 AN ALTERNATE FORMULATION FOR TEMPERATURE AND LACK OF FIT

The class of effects discussed in this section is characterized by the fact that under them the elements to be assembled into a structure have known displacements in their unstressed state. That is, the member displacements do not go to zero with the member forces. For example, a truss bar when subjected to a temperature change has a member displacement or length change of $\alpha \cdot \Delta T \cdot L$ when the force in the member is zero. Here

α — coefficient of thermal expansion,
ΔT — temperature change,
L — length of the member.

To include such effects in the node method simply requires a generalization of Hooke's law, Eq. (3.1), to include a known displacement,

$$F = K(\Delta - D),\qquad (6.1)$$

in which D is a known column matrix. In its generalized form the node method becomes

$$\tilde{N}F = P \quad \rightarrow \quad \tilde{N}K(\Delta - D) = P \quad \rightarrow \quad \tilde{N}KN\delta = P + \tilde{N}KD \quad \rightarrow$$

$$\delta = (\tilde{N}KN)^{-1}(P + \tilde{N}KD).\qquad (6.2)$$

It is of some interest to note finally that the additional term $\tilde{N}KD$ which occurs on the right hand side of Eq. (6.2) is identical with the fictitious forces which occurred under the decomposition described in the previous section.

6.5 EXERCISES

1. Modify the frame Programs P.3 and P.4 to include the effect of uniform member loads.

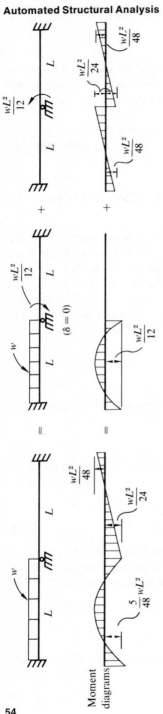

Moment diagrams

Fig. 6.4. Member load.

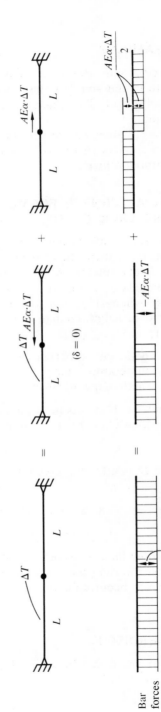

Bar forces

Fig. 6.5. Temperature effects.

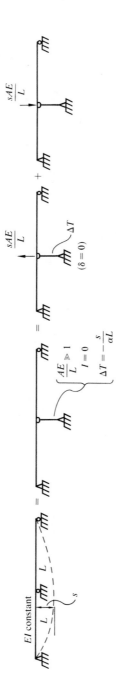

Fig. 6.6. Settlement.

2. Modify the truss Programs P.1 and P.2 to include the effect of temperature. (Note that

$$\tilde{N}KD = \sum_i \tilde{N}_i K_i D_i \qquad \text{and} \qquad \tilde{N}_i K_i D_i = \begin{bmatrix} \vdots \\ n_i K_i D_i \\ \vdots \\ -n_i K_i D_i \\ \vdots \end{bmatrix} \begin{matrix} \\ \text{row } A \\ \\ \text{row } C \\ \\ \end{matrix}$$

using the notation of Chapter 2.)

SEVEN

The Primitive Stiffness Matrix

7.1 INTRODUCTION

Structural analysis, at least as it is presented in this book, has two distinct aspects, the description of individual elements and their assembly into a structure. With the exception of the truss for which the member description is trivial, the first four chapters of this book have avoided any detailed description of members other than straight beams and concerned themselves largely with the problem of assembly.

Through this approach, much of the physics has been put off, and for good reason; it is the most difficult aspect of structures. While the assembly of elements can be handled in a rather routine manner, as this book attemps to show, the meaningfulness of the results obtained is a direct product of the member descriptions used and must stand or fall on their validity.

The behavior of an individual element can be approached on any level of sophistication, mathematically or experimentally. It would surely be inconsistent with the objectives of this book, one of which is simplicity, to deal with the continuum mechanical aspects of member stiffness. In fact, only the use of straight, uniform members is discussed in any detail here. On the surface this would appear to be a considerable restriction.

In this chapter it is shown how a good deal of generality can be achieved through the use of straight uniform members to approximate an arbitrarily curved member. Such an approximation is intuitively appealing, extremely simple, commonly practiced, and in the spirit of the recent developments in finite element methods.

7.2 MEMBER STIFFNESS OBTAINED FROM THE CANTILEVER FLEXIBILITY

For reasons of convenience, it is common to compute the member stiffness by first "fixing" one end of the member and finding its behavior as a cantilever. While this is not a completely general procedure (it is impossible when the member stiffness matrix is singular!), it is frequently quite useful.

Figure 7.1 shows a member which has been "fixed" at its negative end (i.e. $\delta_c = 0$). For convenience in this section it is assumed that the global

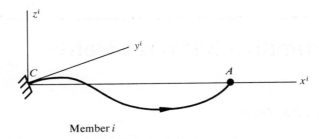

Member i

Fig. 7.1. Member i fixed at its negative end.

coordinate system and the local coordinate system of the member being discussed coincide so that the rotation matrix becomes the identity matrix. Equation (5.2) then becomes

$$\Delta_i = N_i{}^+\delta_A + N_i{}^-\delta_c. \tag{7.1}$$

It is assumed that the behavior of the member as a cantilever is known. This implies that the matrix C_i which satisfies the equation

$$\delta_A = C_i f_i{}^+, \qquad \delta_c = 0 \tag{7.2}$$

is given from which it is desired to find the member stiffness K_i. In Eq. (7.2), $f_i{}^+$ is again the end-of-member force (*see* also Eq. (5.1)). When $\delta_c = 0$, Eq. (7.1) reduces to

$$\Delta_i = N_i{}^+\delta_A. \tag{7.3}$$

Multiplying Eq. (7.2) by $N_i{}^+$ and using Eq. (5.1) it follows that

$$N_i{}^+\delta_A = \Delta_i = N_i{}^+C_i f_i{}^+ = N_i{}^+C_i\tilde{N}_i{}^+F_i, \tag{7.4}$$

and from the definition of the stiffness matrix that

$$K_i = (N_i{}^+C_i\tilde{N}_i{}^+)^{-1}. \tag{7.5}$$

Equation (7.5) can be viewed formally as a transformation from the cantilever flexibility, C_i, to the member stiffness K_i. An example of the use of this technique is given in Section 7.5.

7.3 ADDING FLEXIBILITIES TO APPROXIMATE A CURVED MEMBER

As stated in the introduction to this chapter, arbitrary curved members are treated in this book only as they can be approximated by straight uniform beam segments. Still, a large class of problems can be handled in this manner.

Figure 7.2 shows this procedure schematically. It is assumed that the curved member shown has a given description which allows its approximation by n straight segments. If the cantilever flexibility were known

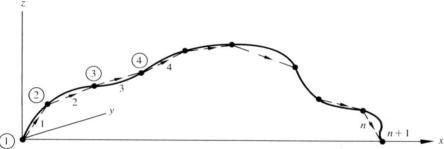

Fig. 7.2. Segmentally straight approximation of a curved beam.

for this member, it would be possible to obtain the member stiffness matrix K_i by using Eq. (7.5). In this section a method for obtaining the cantilever flexibility by adding flexibilities is described. It simply uses the fact that when it is known how to add the flexibilities of members which are connected serially, the cantilever flexibility of the segmentally straight member can be obtained by starting at the fixed end (whose flexibility is zero) and adding segments one at a time.

It is only necessary therefore to consider the situation shown in Fig. 7.3 in which it is implied that the flexibility of point $i-1$ is known when point 1 is fixed. It is desired to obtain the flexibility of point i which results from adding the piece $i-1$ to the partial member. Formally it is assumed that C_{i-1} is given and satisfies the equation

$$\delta_{i-1} = C_{i-1}(\tilde{R}_{i-2}f^+_{i-2}),\tag{7.6}$$

and that it is desired to find C_i which satisfies the equation

$$\delta_i = C_i(\tilde{R}_{i-1}f^+_{i-1}).\tag{7.7}$$

59

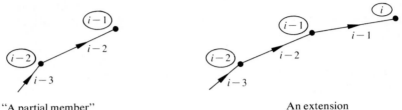

"A partial member"   An extension

Fig. 7.3. Adding flexibilities.

The notation used here is generally consistent with the other sections of the book. For example, $(\tilde{R}_{i-1}f^{+}_{i-1})$ is the end of member force on the positive end of member $i-1$ in the global coordinate system. In Eqs. (7.6 and 7.7) the parentheses have been added for clarity.

To find C_i, it is convenient to use the equilibrium equation of joint $i-1$,

$$-\tilde{R}_{i-2}f^{+}_{i-2} = \tilde{R}_{i-1}f^{-}_{i-1},\tag{7.8}$$

Eqs. (5.1),

$$f^{+}_i = \tilde{N}_i^{+}F_i, \qquad f^{-}_i = \tilde{N}_i^{-}F_i,$$

and Eq. (5.2),

$$\Delta_i = N_i^{+}R_i\delta_A + N_i^{-}R_i\delta_c \quad \rightarrow \quad \delta_c = \tilde{R}_i(N_i^{-})^{-1}(\Delta_i - N_i^{+}R_i\delta_A).\tag{7.9}$$

Starting from Eq. (7.6) and using Eq. (7.8) it follows that

$$\delta_{i-1} = -C_{i-1}(\tilde{R}_{i-1}f^{-}_{i-1}),$$

but using Eqs. (7.9 and 5.1) leads to

$$\tilde{R}_{i-1}(N^{-}_{i-1})^{-1}(\Delta_{i-1} - N^{+}_{i-1}R_{i-1}\delta_i) = -C_{i-1}(\tilde{R}_{i-1}\tilde{N}^{-}_{i-1}F_{i-1}).$$

Applying Hooke's law, $F_i = K_i\Delta_i$, and then the first of Eqs. (5.1) again results in

$$\tilde{R}_{i-1}(N^{-}_{i-1})^{-1}(K^{-1}_{i-1}F_{i-1} - N^{+}_{i-1}R_{i-1}\delta_i) = -C_{i-1}(\tilde{R}_{i-1}\tilde{N}^{-}_{i-1}F_{i-1})$$

$$[\tilde{R}_{i-1}(N^{-}_{i-1})^{-1}K^{-1}_{i-1} + C_{i-1}\tilde{R}_{i-1}\tilde{N}^{-}_{i-1}]F_{i-1} = \tilde{R}_{i-1}(N^{-}_{i-1})^{-1}N^{+}_{i-1}R_{i-1}\delta_i$$

$$\tilde{R}_{i-1}(N^{+}_{i-1})^{-1}[K^{-1}_{i-1} + N^{-}_{i-1}R_{i-1}C_{i-1}\tilde{R}_{i-1}\tilde{N}^{-}_{i-1}](\tilde{N}^{+}_{i-1})^{-1}R_{i-1}(\tilde{R}_{i-1}f^{+}_{i-1}) = \delta_i$$

or simply

$$C_i = \tilde{R}_{i-1}(N^{+}_{i-1})^{-1}[K^{-1}_{i-1} + N^{-}_{i-1}R_{i-1}C_{i-1}\tilde{R}_{i-1}\tilde{N}^{-}_{i-1}](\tilde{N}^{+}_{i-1})^{-1}R_{i-1}.\tag{7.10}$$

This is the desired result. Starting at the fixed support, point 1, at which $C_1 = 0$, it is now possible to add pieces until the entire curved member has been approximated.

It may be noted finally that while this chapter has been largely concerned with the approximation of curved members by straight members, so far as the formal aspects of Eq. (7.10) are concerned, K_i is arbitrary and it is possible to think of approximating arbitrarily curved members by, e.g., circular segments or any other convenient pieces whose member descriptions are available.

Program P.7 illustrates the use of Eq. (7.10) in the approximation of a plane curved member by straight uniform segments.

7.4 SINGULAR STIFFNESS MATRICES

The node method for frames as it is presented here has a tendency to appear more restricted than it really is. While it is perhaps not obvious, such problems as hinges or releases, for example, in general can be handled very simply by including their effects in member stiffness matrices. Some applications of this method are indicated in Fig. 7.4. When this is done, the member stiffness matrix becomes singular.

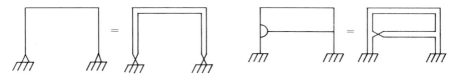

Fig. 7.4. Releases included in the member stiffness matrix.

A useful example of a singular stiffness matrix is the case of the uniform straight beam with a hinge at some arbitrary point αL as shown in Fig. 7.5.

Using the results of Appendix 3 it is possible to obtain the angle ψ, the rotation due to a unit moment, as shown in Fig. 7.6. Summing steps 1 and 2 gives

$$\psi = \frac{L}{3EI} \frac{(1-\alpha)^3 + \alpha^3}{\alpha^2} \tag{7.11}$$

from which it follows that the stiffness, (i.e. the moment required to produce a unit rotation), is

$$K_{33} = \frac{3EI}{L} \frac{\alpha^2}{3\alpha(\alpha-1)+1} = \frac{1}{\psi} \tag{7.12}$$

The stiffness of the other end is obtained by replacing α by $1-\alpha$ in Eq. (7.12) to obtain

$$K_{22} = \frac{3EI}{L} \frac{(1-\alpha)^2}{3\alpha(\alpha-1)+1}. \tag{7.13}$$

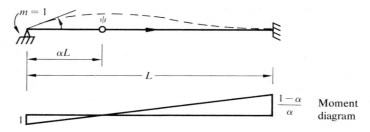

Fig. 7.5. Beam with a hinge.

Finally, the stiffness matrix for the entire member is (the subscripts are omitted)

$$K = \begin{bmatrix} \dfrac{AE}{L} & 0 & 0 \\[2ex] 0 & \dfrac{3EI}{L}\dfrac{(1-\alpha)^2}{3\alpha(\alpha-1)+1} & \dfrac{3EI}{L}\dfrac{\alpha^2}{(1-\alpha)^3+\alpha^2}\dfrac{1-\alpha}{\alpha} \\[2ex] 0 & \dfrac{3EI}{L}\dfrac{(1-\alpha)^2}{3\alpha(\alpha-1)+1}\dfrac{\alpha}{(1-\alpha)} & \dfrac{3EI}{L}\dfrac{\alpha^2}{3\alpha(\alpha-1)+1} \end{bmatrix}. \qquad (7.14)$$

7.5 THE EFFECT OF SHEAR

As a final example of the generality available under the methods of this book, this section describes how shear effects can be included easily in the stiffness matrix for plane, straight uniform beams. In order to do so it is convenient to think of members connected at joints by rigid blocks as shown in Fig. 7.7. The point of this figure is that while ordinary beam theory assumes that "plane sections remain plane" and that cross sections remain perpendicular to the beam centerline under deflection, beams can actually be assembled so that their end cross sections rotate the same amount, and it is not necessary for their centerlines to remain perpendicular to their cross sections.

Figure 7.8 shows the shear deformation of a beam element. The shearing angle γ is usually written

$$\gamma = \frac{V}{kAG} \qquad (7.15)$$

The coefficient k is commonly determined using energy methods (*see* Timoshenko) but for the case of wide flange beams it can be taken to be unity in Eq. (7.15) when A is taken to be the area of the web. In the remainder of this section a modified member stiffness matrix which

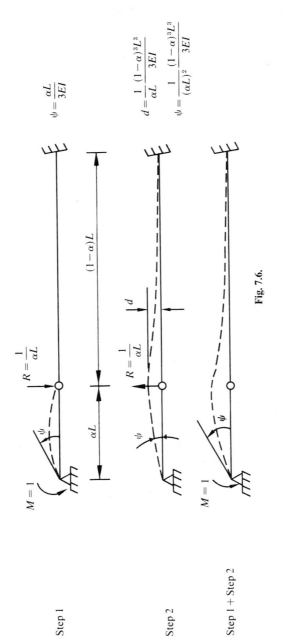

Step 1

$R = \dfrac{1}{\alpha L}$

$M = 1$

$\psi = \dfrac{\alpha L}{3EI}$

αL

$(1 - \alpha)L$

Step 2

$R = \dfrac{1}{\alpha L}$

d

$d = \dfrac{1}{\alpha L}\dfrac{(1-\alpha)^3 L^3}{3EI}$

$\psi = \dfrac{1}{(\alpha L)^2}\dfrac{(1-\alpha)^3 L^3}{3EI}$

Step 1 + Step 2

$M = 1$

Fig. 7.6.

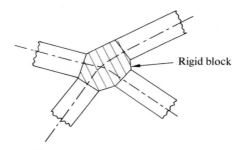

Fig. 7.7. Members connected through a rigid block.

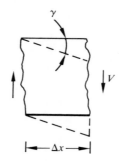

Fig. 7.8. Shear deformation of a beam element.

reduces to the stiffness matrix given in Chapter 4 as $k \to \infty$ is derived by first deriving its cantilever flexibility and then using Eq. (7.5) to obtain the member stiffness.

Consider the beam shown in Fig. 7.9. Using the results of Appendix 3 and Eq. (7.15), it has the following characteristics under load:

| | Free End Displacement | | |
Load	Horizontal Component	Vertical Component	Rotation
Unit horizontal load	$\dfrac{L}{AE}$	0	0
Unit vertical load	0	$\dfrac{L^3}{3EI} + \dfrac{L}{kAG}$	$\dfrac{L^2}{2EI}$
Unit moment	0	$\dfrac{L^2}{2EI}$	$\dfrac{L}{EI}$

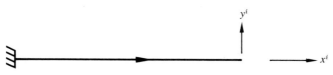

Fig. 7.9. A plane beam.

The cantilever flexibility is therefore

$$C_i = \begin{bmatrix} \dfrac{L}{AE} & 0 & 0 \\[2ex] 0 & \dfrac{L^3}{3EI} + \dfrac{L}{kAG} & \dfrac{L^2}{2EI} \\[2ex] 0 & \dfrac{L^2}{2EI} & \dfrac{L}{EI} \end{bmatrix}$$

and the member stiffness.

$$K_i = (N_i^+ C_i \tilde{N}_i^+)^{-1} = \begin{bmatrix} \dfrac{AE}{L} & 0 & 0 \\[3ex] 0 & \dfrac{\dfrac{L}{3EI} + \dfrac{1}{AGkL}}{\nabla} & \dfrac{\dfrac{L}{6EI} - \dfrac{1}{AGkL}}{\nabla} \\[4ex] 0 & \dfrac{\dfrac{L}{6EI} - \dfrac{1}{AGkL}}{\nabla} & \dfrac{\dfrac{L}{3EI} + \dfrac{1}{AGkL}}{\nabla} \end{bmatrix}$$

in which

$$\nabla = \frac{L^2}{12E^2I^2}\left(1 + \frac{12EI}{AGkL^2}\right)$$

7.6 EXERCISES

1. Modify the frame Programs P.3 and P.4 to include the effects of member hinges and shear.
2. Modify the frame Programs P.3 and P.4 to use Program P.7 as a subroutine when dealing with curved members.

EIGHT

The Mesh Method

8.1 INTRODUCTION

The force method, an alternative form of the mesh method to which this chapter is devoted, was until recently the most popular method of structural analysis. While the decline of the force method can only be attributed to programming difficulties, it retains certain important advantages which on occasion overshadow these difficulties. The principal advantage of the mesh method is its ability to tolerate rigid elements which cause numerical difficulties (ill-conditioning) in the node method; this situation can be compared with the ease by which hinges (*see* Chapter 7) which cause problems in the mesh method, can be included in the node method.

The mesh method for *networks* (*see* Chapter 1) is purely topological and quite elegant. It is based upon the idea that rather than writing equilibrium equations in terms of node quantities, it is possible to combine solutions (mesh flows), each of which satisfies equilibrium, to satisfy the mesh law.

The situation for structures is basically the same as that of networks but much more complicated. In structures it is not sufficient to know connectivity, geometry must be considered and it is here that the difficulties lie. The magnitude of these difficulties becomes apparent when the generalization of the branch-mesh matrix is generated in this chapter. While N, the generalization of the branch-node matrix has occasionally become complicated, it has nevertheless always been possible to give a rather simple description of it. On the other hand, the generalization of the branch-mesh matrix which is used in Program P.5 is more complicated by an order of magnitude and is prescribed in a constructive manner.

In the next section a rather formal derivation of the mesh method is given in terms of the node method. There, ideas of the degree of statical indeterminacy and compatibility are introduced in the by now familiar notation of the node method. This section is followed by an attempt to discuss simply the physics of the mesh method. Finally, a general statement of the mesh method is presented and followed by exercises.

8.2 THE MESH METHOD DERIVED FROM THE NODE METHOD

In this section the equations of the node method are rewritten to obtain a formal statement of the mesh method. Recalling the equations of the node method,

$$\tilde{N}F = P \qquad F = K\Delta \qquad \Delta = N\delta,$$

for a stable structure they can always be partitioned into

$$[\tilde{N}_T \tilde{N}_L]\begin{bmatrix} F_T \\ F_L \end{bmatrix} = P,$$

$$\begin{bmatrix} F_T \\ F_L \end{bmatrix} = \begin{bmatrix} K_T & 0 \\ 0 & K_L \end{bmatrix}\begin{bmatrix} \Delta_T \\ \Delta_L \end{bmatrix} \qquad (8.1)$$

$$\begin{bmatrix} \Delta_T \\ \Delta_L \end{bmatrix} = \begin{bmatrix} N_T \\ N_L \end{bmatrix}\delta$$

after perhaps an interchange of rows, so that the matrix N_T is square and non-singular (i.e. N_T^{-1} exists). The elements of F_L constitute what is usually referred to as a set of redundants. In fact the number of independent elements in F_L is called the *degree of statical indeterminacy*. This is made more clear by writing the equilibrium equations as

$$F_T = \tilde{N}_T^{-1}(P - \tilde{N}_L F_L) \qquad (8.2)$$

from which it follows that equilibrium is satisfied for arbitrary F_L provided that F_T is computed using Eq. (8.2). From the last of Eqs. (8.1) it follows that

$$\delta = N_T^{-1}\Delta_T \qquad \text{and} \qquad \Delta_L = N_L\delta = N_L N_T^{-1}\Delta_T. \qquad (8.3)$$

The last of Eq. (8.3) relates the member displacements Δ_L associated with redundant bars to the member displacements Δ_T of the other bars. This *compatibility equation* simply states that once Δ_T has been specified, Δ_L is also determined if the bars are to fit together.

The compatibility equation is the basis for the mesh method which follows directly once the member displacements are written in terms of the matrix F_L, and the joint load matrix P. This is achieved simply by writing

$$\Delta_L = N_L N_T^{-1} \Delta_T \quad \rightarrow \quad K_L^{-1} F_L = N_L N_T^{-1} K_T^{-1} \tilde{N}_T^{-1} (P - \tilde{N}_L F_L)$$

$$\text{(using Eq. (8.1))}$$

or

$$(K_L^{-1} + N_L N_T^{-1} K_T^{-1} \tilde{N}_T^{-1} \tilde{N}_L) F_L = N_L N_T^{-1} K_T^{-1} \tilde{N}_T^{-1} P. \qquad (8.4)$$

Equation (8.4) is the equation of the mesh method; however, the form indicated is not at all convenient for computation since it requires the inversion of N_T. In a later section a more convenient form is developed and the remainder of this section is devoted to a modified form of Eq. (8.4) motivated by the topological network problem. After the topological branch-mesh matrix, let

$$C = \begin{bmatrix} C_T \\ C_L \end{bmatrix} = \begin{bmatrix} -N_T^{-1} & N_L \\ I & \end{bmatrix}, \qquad (8.5)$$

in terms of which the compatibility equation, Eq. (8.3), becomes

$$\tilde{C}\Delta = 0. \qquad (8.6)$$

Following networks a little further, F can be decomposed into

$$F = \mathscr{F} + f, \qquad (8.7)$$

where $\mathscr{F}$ is assumed to be known and to have the property

$$\tilde{N}\mathscr{F} = P, \qquad (8.8)$$

so that

$$\tilde{N}f = 0 \qquad (8.9)$$

and the problem (in terms of f) has been reduced to one for which there are no applied node forces. The decomposition, Eq. (8.7), is not unique and can be done in many ways, one of which is to let

$$\mathscr{F} = \Theta P, \quad \text{where} \quad \Theta = \begin{bmatrix} N_T^{-1} \\ 0 \end{bmatrix}, \qquad (8.10)$$

from which it follows that

$$f = CF_L. \qquad (8.11)$$

In terms of these variables it follows that

$$f = CF_L = F - \mathscr{F} = F - \Theta P \qquad (8.12)$$

or

$$F = CF_L + \Theta P. \qquad (8.13)$$

69

Formally the equations of the mesh method are then

$$\tilde{C}\Delta = 0 \qquad \text{(mesh law)}$$
$$\Delta = K^{-1}F \qquad \text{(Hooke's law)} \qquad (8.14)$$
$$F = CF_L + \Theta P \qquad \text{(branch force-mesh force relationship)}$$

from which it follows that

$$\tilde{C}\Delta = 0 \quad \rightarrow \quad \tilde{C}K^{-1}F = 0 \quad \rightarrow \quad \tilde{C}K^{-1}(CF_L + \Theta P) = 0, \quad (8.15)$$

or

$$F_L = -(\tilde{C}K^{-1}C)^{-1}\tilde{C}K^{-1}\Theta P, \qquad (8.16)$$

which is identical to Eq. (8.4).

8.3 THE MESH LAW

In Section 8.2 the mesh method has been derived directly from the node method. It is however most common to postulate a "mesh law" for a system from which the mesh method can be obtained directly. That is done in this chapter for skeletal structures which have well behaved primitive flexibilities (i.e. $|K_i^{-1}| \neq 0$).

The mesh law in this case might be stated, "The relative displacement of two adjacent points as computed by moving around any closed loop must be zero." The idea here is that given member displacements, it is possible to "fix" a reference joint and then go from joint to joint through the structure computing displacements. In particular, the computed displacement of the reference point as determined by moving around any path must be zero. For stable structures the number of independent conditions which can be obtained in this fashion equals the degree of statical indeterminacy of the structure.

Figure 8.1 shows a simple structure to which this process is applied. It is convenient to use Eq. (5.2) in the form

$$\Delta_i = \eta_i^+ \delta_A + \eta_i^- \delta_c. \qquad (8.17)$$

Starting at joint 1,

1. Let $\quad \delta_1 = 0$
2. Compute $\delta_2 = (\eta_1^+)^{-1}\Delta_1$
3. Compute $\delta_3 = (\eta_2^+)^{-1}[\Delta_2 - \eta_2^- \delta_2]$
   $$= (\eta_2^+)^{-1}[\Delta_2 - \eta_2^-(\eta_1^+)^{-1}\Delta_1]$$
4. Compute $\delta_1 = (\eta_3^-)^{-1}[\Delta_3 - \eta_3^+ \delta_3]$
   $$= (\eta_3^-)^{-1}[\Delta_3 - \eta_3^+(\eta_2^+)^{-1}\{\Delta_2 - \eta_2^-(\eta_1^+)^{-1}\Delta_1\}]$$
   $$= 0$$

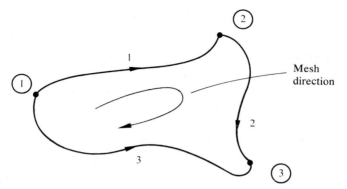

Fig. 8.1. A single mesh frame.

Collecting terms gives the compatibility equation

$$(\eta_3{}^-)^{-1}\eta_3{}^+(\eta_2{}^+)^{-1}\eta_2{}^-(\eta_1{}^+)^{-1}\Delta_1 - (\eta_3{}^-)^{-1}\eta_3{}^+(\eta_2{}^+)^{-1}\Delta_2 + (\eta_3{}^-)^{-1}\Delta_3 = 0.$$

$$(8.18)$$

An equation like this can be written for any closed loop (mesh) within a structure. It is of some interest to compare this equation to Eq. (1.11) for networks.

Meshes also play an important role with regard to force systems. In particular, it is possible to construct force systems which satisfy equilibrium for an unloaded structure ($P = 0$) by selecting arbitrarily the force in one member of a mesh and then proceeding from member to member around the mesh computing member forces. It is convenient to use Fig. 8.1 again to demonstrate this procedure. But first write Eq. (5.1) in the form,

$$\tilde{R}_i f_i{}^+ = \tilde{R}_i \tilde{N}_i{}^+ F_i = \tilde{\eta}_i{}^+ F_i \qquad \text{and} \qquad \tilde{R}_i f_i{}^- = \tilde{R}_i \tilde{N}_i{}^- \tilde{F}_i = \tilde{\eta}_i{}^- F_i. \qquad (8.19)$$

Starting with member 1,

1. Let F_1 be arbitrary
2. Compute $\tilde{R}_1 f_1{}^+ = \tilde{\eta}_1{}^+ F_1$
3. But $\quad \tilde{R}_1 f_1{}^+ = -\tilde{R}_2 f_2{}^-$ (equilibrium)
4. Therefore $\quad F_2 = (\tilde{\eta}_2{}^-)^{-1}(\tilde{R}_2 f_2{}^-) = -(\tilde{\eta}_2{}^-)^{-1}\tilde{\eta}_1{}^+ \tilde{F}_1$
5. Compute $\tilde{R}_2 f_2{}^+ = \tilde{\eta}_2{}^+ F_2$
6. But $\quad \tilde{R}_2 f_2{}^+ = \tilde{R}_3 f_3{}^+$ (equilibrium)
7. Therefore $\quad F_3 = (\tilde{\eta}_3{}^+)^{-1}(\tilde{R}_3 f_3{}^+) = (\tilde{\eta}_3{}^+)^{-1}\tilde{\eta}_2{}^+(\tilde{\eta}_2{}^-)^{-1}\tilde{\eta}_1{}^+ F_1$

Note that if Eq. (8.18), the compatibility equation, is cleared so that the coefficient of Δ_1 is I, the coefficients above which relate F_1 to the member

71

forces become identical with the transpose of the coefficients which occur in the compatibility equation.

It should be clear now that a set of member forces which satisfy equilibrium with respect to the loaded structure can be found by first reducing the structure to a tree and then moving inward from the tree tips computing member forces as has been done in the preceding paragraph but using the appropriate nonhomogeneous equilibrium equations.

8.4 AN ALTERNATIVE FORMULATION

There is one aspect of the formulation of Section 8.2 which is not completely general. It is the fact that the lower part of the matrix C is taken to be the identity matrix and it is associated with the fact that it is possible to identify each element of F_L with a branch force. (In graph theory this is equivalent to identifying meshes through tree links.) In this section a formulation is presented which does not include such a restriction on C and which is practically motivated by numerical considerations. Briefly, the number of non-zero terms in the system matrix of the mesh method can be reduced by using the more general form of C; this more general form is used in Program P.5.

In the alternative formulation Eq. (8.13) is replaced by

$$F = \mathscr{C}F_M + \bar{F}. \tag{8.20}$$

Here F is any set of forces which satisfy equilibrium,

$$\tilde{N}\bar{F} = P, \tag{8.21}$$

and $\mathscr{C}F_M$ represent independent force systems which satisfy equilibrium with regard to the unloaded structure. It is assumed that the number of elements in the "mesh force matrix" F_M equals the degree of statical indeterminacy of the structure. Defining $\mathscr{C}$ through the use of Eq. (8.20) rather than using Eq. (8.13) is equivalent to allowing *any* mesh description for a graph rather than being restricted to link-tree description of meshes. In view of equilibrium, it follows that

$$\tilde{N}(\mathscr{C}F_M) = 0 \quad \rightarrow \quad \tilde{N}\mathscr{C} = 0, \tag{8.22}$$

and the mesh method follows

$$\tilde{\mathscr{C}}\Delta = 0 \quad \rightarrow \quad \tilde{\mathscr{C}}K^{-1}F = 0 \quad \rightarrow \quad \tilde{\mathscr{C}}K^{-1}(\mathscr{C}F_M + \bar{F}) = 0$$

or

$$\tilde{\mathscr{C}}K^{-1}\mathscr{C}F_M = -\tilde{\mathscr{C}}K^{-1}\bar{F} \tag{8.23}$$

as before.

In conclusion it is perhaps worthwhile to repeat an earlier remark. It is only possible to relate the matrices F_L and F_M to topological quantities (meshes) for structures without hinges (in general releases). For trusses and other structures which contain releases, the mesh method must be approached by examining the rank of the matrix N as was done in Section 8.2.

8.5 EXERCISES

1. Modify the Program P.5 to generate its own mesh descriptions rather than reading them in as input.
2. Modify Program P.5 to include the effects of variable moment of inertia, member load, and temperature.

NINE

Miscellaneous Theorems

In the context of this book, several of the classical structural theorems take on an especially simple form and are collected in this chapter.

For all problems considered in this book, the system matrix has been symmetric. For example, in the node method

$$(\tilde{N}KN)\delta = P, \tag{9.1}$$

the system matrix $\tilde{N}KN$ is symmetric since in general

$$(\widetilde{ABC}) = \tilde{C}\tilde{B}\tilde{A}, \tag{9.2}$$

from which it follows that

$$(\widetilde{\tilde{N}KN}) = \tilde{N}KN, \tag{9.3}$$

if K is symmetric, (i.e. $\tilde{K} = K$). For all the cases considered, K has been symmetric. Rewriting Eq. (9.1) as

$$C\delta = P, \tag{9.4}$$

since C is symmetric C^{-1} is also symmetric. From these symmetries follow

MAXWELL'S LAW. *The displacement i due to a unit force j equals the displacement j due to a unit force i.* (There is a second form of this theorem which is obtained by interchanging the words force and displacement in the theorem stated.)

Proof: $C_{ij} = C_{ji}$.

As stated, this theorem is a little out of the spirit of this book in that when it refers to a displacement or force it means a single scalar force or displacement component.

Given two different sets of loads P^1 and P^2 and their associated displacements δ^1 and δ^2 produced on the same structure. To this situation applies

BETTI'S LAW. $\tilde{\delta}^1 P^2 = \tilde{\delta}^2 P^1$.

Proof: Eq. (9.4) applies to both loads to give

$$C\delta^1 = P^1 \qquad \text{and} \qquad C\delta^2 = P^2.$$

Multiplying by $\tilde{\delta}^2$ and $\tilde{\delta}^1$ respectively results in

$$\tilde{\delta}^2 C\delta^1 = \tilde{\delta}^2 P^1 \qquad \text{and} \qquad \tilde{\delta}^1 C\delta^2 = \tilde{\delta}^1 P^2.$$

But $\tilde{\delta}^2 C\delta^1 = \tilde{\delta}^1 C\delta^2$, since C is symmetric and a scalar is equal to its transpose.

In Chapter 3 it was explained that the formulation of the node method implies that

$$W = \tfrac{1}{2}\tilde{P}\delta = \tfrac{1}{2}\tilde{F}\Delta = E, \tag{9.5}$$

from which it follows directly

CASTIGLIANO'S THEOREM. $\partial E/\partial(\delta)_I = (P)_I$.

Proof: $E = W = \tfrac{1}{2}\tilde{\delta}C\delta$. By direct differentiation it can be shown that $\partial E/\partial(\delta)_I = (C\delta)_I = (P)_I$.

Note that again there is a second form of the theorem which states that $\partial E/\partial(P)_I = (\delta)_I$ and which can also be proved by direct differentiation.

VIRTUAL WORK.   *Given any force system which satisfies equilibrium,*

$$\tilde{N}F^1 = P^1 \tag{9.6}$$

and any "conformable" set of displacements δ, then

$$\tilde{\Delta}F^1 = \tilde{\delta}P^1. \tag{9.7}$$

Proof: Eq. (9.7) follows directly by multiplying Eq. (9.6) on the left by $\tilde{\delta}$,

$$\tilde{\delta}\tilde{N}F^1 = \tilde{\delta}P^1, \tag{9.8}$$

and interpreting $N\delta = \Delta$ as a set of member displacements. δ is "conformable" if N and δ are of proper dimensions so that $N\delta$ has meaning.

The generality of Eq. (9.7) lies in the fact that the displacements and forces do not have to occur in the same structure. This fact has been long recognized by engineers who have used statically determinate "virtual structures" when computing displacements in statically indeterminate structures.

Finally, geometric stability is easily discussed in the notation of this book. If a structure is (geometrically) unstable, there exists a displaced configuration $\delta \neq 0$ for which no bar forces are generated. The implication is that at least part of the structure is behaving like a mechanism under this displacement.

Let the matrix N have m columns.

THEOREM. *If a structure is unstable, the rank of KN must be less than m.*

Proof: By definition, if a structure is unstable, there exists a $\delta \neq 0$ and an associated $F = 0$.

Since

$$F = 0 = K\Delta = KN\delta, \tag{9.9}$$

if the theorem were not true it would be possible to partition Eq. (9.9) after a re-arrangement of rows

$$\left[-\frac{F^1}{F^2} \right] = \left[\frac{(KN)^1}{(KN)^2} \right] \delta,$$

so that $(KN)^1$ would have an inverse and F^1 would not in general be zero. This is the desired contradiction.

Note that in general stability depends upon K. In cases such as trusses in which none of the bar areas is zero and which therefore have non-singular primitive stiffness matrices, stability depends only upon the matrix N.

TEN

Kron's Methods

10.1 INTRODUCTION

With a little experience it becomes clear that one of the major problems facing the structural analyst is how to reduce the computational effort (cost) of solutions which is closely related to some very practical problems of describing structures and information retrieval. Large systems are addressed directly in Appendix A.5 but in this chapter some related work of the late Gabriel Kron is discussed as it applies to structures.

While the intention here is not to imply that Kron was the only worker in these fields, it seems clear now that he was far ahead of his time and, in fact, it still remains to investigate the generality and usefulness of some of his techniques.

10.2 TEARING AND INTERCONNECTING

Kron is best known for his method of "tearing and interconnecting" in which rather than solving a structure directly, pieces of it are first removed (the structure is torn) and the reduced structure is solved. Then the effect of the missing pieces is introduced (the structure is interconnected). Under the proper conditions, solution by tearing and interconnecting can produce a phenomenal cost saving.

Tearing and interconnection is most directly explained in terms of the method of modified matrices based upon the formula

$$(B - US\tilde{V})^{-1} = B^{-1} + B^{-1}U(S^{-1} - \tilde{V}B^{-1}U)^{-1}\tilde{V}B^{-1}, \qquad (10.1)$$

which can be verified directly by starting with the identity

$$US[(S^{-1} - \tilde{V}B^{-1}U)(S^{-1} - \tilde{V}B^{-1}U)^{-1} - I]\tilde{V}B = 0,$$

adding I to each side of the equation and expanding,

$$I + U(S^{-1} - \tilde{V}B^{-1}U)^{-1}\tilde{V}B^{-1} - US\tilde{V}B^{-1}$$
$$- US\tilde{V}B^{-1}U(S^{-1} - \tilde{V}B^{-1}U)^{-1}\tilde{V}B^{-1} = I,$$

producing finally

$$(B - US\tilde{V})[B^{-1} + B^{-1}U(S^{-1} - \tilde{V}B^{-1}U)^{-1}\tilde{V}B^{-1}] = I.$$

Equation (10.1) is usually applied to the situation in which the behavior of a system, B^{-1}, is known, given the system matrix B and it is desired to find the effect of a change in B, $-US\tilde{V}$, on the inverse system matrix B^{-1}.

The implications of Eq. (10.1) are quite broad and will be discussed here largely through examples, but it is first interesting to note the similarity in form between Eq. (10.1) and Eq. (3.6).

Figure 10.1 shows a very simple plane truss which is "torn" into two identical independent structures by removing bar 5. For simplicity, assume that for each bar in this problem $K_i = A_iE/L_i = 1$. The system matrix for the torn structure must have the form

$$(\tilde{N}KN)_{\text{torn}} = \begin{bmatrix} A & 0 \\ 0 & A \end{bmatrix}, \tag{10.2}$$

since the two pieces are identical and independent.

Using the decomposition procedure described in Chapter 3, the matrix A is found to be

$$A = \frac{1}{\sqrt{2}}\begin{bmatrix} 1 \\ 1 \end{bmatrix}[1 \quad 1]\frac{1}{\sqrt{2}} + \frac{1}{\sqrt{2}}\begin{bmatrix} -1 \\ 1 \end{bmatrix}[-1 \quad 1]\frac{1}{\sqrt{2}}$$

$$= \begin{bmatrix} 1 & 0 \\ 0 & 1 \end{bmatrix}$$

and the system matrix in Eq. (10.2) is seen to be the identity matrix. Referring to Eq. (3.6) again, the system matrix for the original system may be written

$$\tilde{N}KN = \sum_{i=1}^{4} (\tilde{N}_iK_iN_i) + \tilde{N}_5K_5N_5$$

$$= (\tilde{N}KN)_{\text{torn}} + \tilde{N}_5K_5N_5 \tag{10.3}$$

where

$$N_5 = [-[1,0] \quad [1,0]] \quad \text{and} \quad K_5 = 1.$$

Equation (10.1) can now be used to find $(\tilde{N}KN)^{-1}$ by identifying

$$\begin{aligned} B &= (\tilde{N}KN)_{\text{torn}} = I \\ S &= -1 \\ V &= U = \tilde{N}_5 \end{aligned} \tag{10.4}$$

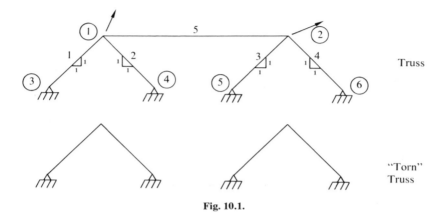

Fig. 10.1.

after which it follows from Eq. (10.1) that

$$(\tilde{N}KN)^{-1} = I + I\tilde{N}_5(-1 \quad -[-1 \quad 0 \quad 1 \quad 0]\begin{bmatrix} -1 \\ 0 \\ 1 \\ 0 \end{bmatrix})^{-1}N_5I$$

$$= \begin{bmatrix} \frac{2}{3} & 0 & \frac{1}{3} & 0 \\ 0 & 1 & 0 & 0 \\ \frac{1}{3} & 0 & \frac{2}{3} & 0 \\ 0 & 0 & 0 & 1 \end{bmatrix}$$

Another application of Eq. (10.1) called "doubling" is shown schematically in Fig. 10.2. The idea here is to tear a structure into a number of independent identical pieces and interconnect them using a number of steps, each of which doubles the size of the torn structure, until the original structure is achieved.

Anticipating Appendix A.5 a little, it may be noted that one of the shortcomings of tearing is that the inverse system matrix, or at least a large part of it, is constructed while Appendix A.5 shows that this is generally inefficient. It may also be noted that efficient application of tearing requires structures which are highly "symmetric," a situation which is more common in networks than structures.

10.3 K-PARTITIONING

Given a system (structure) in the form

$$A\delta = P, \tag{10.5}$$

81

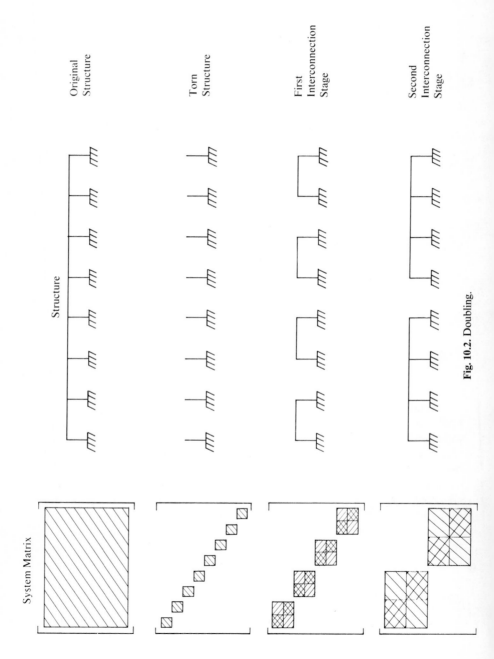

Fig. 10.2. Doubling.

rather than solving it as

$$\delta = A^{-1}P, \tag{10.6}$$

the system can be partitioned into

$$\begin{bmatrix} A_1 & A_2 \\ A_3 & A_4 \end{bmatrix} \begin{bmatrix} \delta_1 \\ \delta_2 \end{bmatrix} = \begin{bmatrix} P_1 \\ P_2 \end{bmatrix}, \tag{10.7}$$

and solved as

$$\delta_2 = (A_4 - A_3 A_1^{-1} A_2)^{-1} (P_2 - A_3 A_1^{-1} P_1), \tag{10.8}$$

and

$$\delta_1 = A_1^{-1}(P_1 - A_2\delta_2).$$

Obviously, this is just a matrix form of Gaussian elimination.

As with all methods for solution, the question of generality or how to proceed in the case of an arbitrary system must be considered. For example, into how many pieces should the system matrix be partitioned? In what order should the elimination be performed?... Some of these problems are dealt with in Appendix A.5, some remain open questions.

10.4 METHOD OF SUBSTRUCTURES

It is sometimes convenient to subdivide a structure into its component parts for analysis. While this is not always easy to motivate from the point of view of computational effort, for practical reasons it may be desirable to do so in order to. e.g., allow design groups to proceed almost independently on their respective parts. In this section it will be shown that the mechanics of using substructures is simply the mechanics of K-partitioning or matrix Gaussian elimination.

Figure 10.3 shows schematically an aircraft which has been subdivided

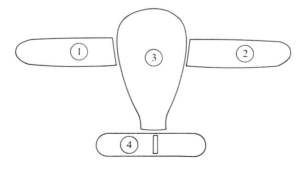

Fig. 10.3. Method of substructures.

into its wings, fuselage, and tail assembly. Its system matrix obviously has the form

$$
\begin{bmatrix}
A_{11} & 0 & 0 & A_{14} \\
0 & A_{22} & 0 & A_{24} \\
0 & 0 & A_{33} & A_{34} \\
A_{41} & A_{42} & A_{43} & A_{44}
\end{bmatrix}
\begin{bmatrix}
\delta_1 \\ \delta_2 \\ \delta_3 \\ \delta_4
\end{bmatrix}
=
\begin{bmatrix}
P_1 \\ P_2 \\ P_3 \\ P_4
\end{bmatrix}.
\tag{10.9}
$$

In the hope that the interactions between the substructures are small, the wing, unit 1, might, e.g., be designed under the assumption that δ_4 is zero or using some approximate values of δ_4 if they are available.

It is common that the matrices shown in Eq. (10.8) are themselves sparse and an interesting exercise to observe the effect of using additional substructures on the form of the system matrix. Finally, it is one of the implicit advantages of using substructures that zero elements in the system matrix show up clearly and are of course not operated upon during the solution.

References

Azar, J. J., *Matrix Structural Analysis in Engineering Mechanics*, Pergamon Press, New York, 1971.

Berge, Claude, *The Theory of Graphs and Its Applications,* John Wiley, New York, 1962.

Branin, Franklin H., *"The Relation Between Kron's Method and the Classical Methods of Network Analysis"*, IRE, Wescon Convention Record, 1959, Part 2.

Kron, Gabriel, *Diakoptics*. Macdonald, London, 1963. (References to most of Kron's work can be found in this collection.)

Maxwell, Lee M. and Reed, Myril B., *The Theory of Graphs: A Basis for Network Theory*, Pergamon Press, New York, 1971.

Norris, Charles and Wilbur, John, *Elementary Structural Analysis*, McGraw Hill, 1960.

Ore, Oystein., *Theory of Graphs*, American Mathematical Society Colloquium Publications. Vol. 38, Providence, R. I., 1962.

Timoshenko, S., *Strength of Materials*, Part II, 2nd Ed., D. Van Nostrand, New York, 1941.

Zienkiewicz, O. C., *The Finite Element Method in Structural and Continuum Mechanics*, McGraw Hill, New York, 1967.

APPENDIX A.1

Matrices and Vectors

A.1.1 MATRIX NOTATION

It is the attempt of this book to use the most simple matrix notation possible. In this vein, no special notation is used for square matrices, diagonal matrices, etc., which has the result of placing an additional burden on the reader. In general he must understand the context of the material to be able to follow the notation.

Matrix notation is probably most easily motivated through a system of simultaneous, linear, algebraic equations,

$$a_{11}x_1 + a_{12}x_2 + \cdots + a_{1n}x_n = b_1$$
$$a_{21}x_1 + a_{22}x_2 + \cdots + a_{2n}x_n = b_2$$
$$\vdots$$
$$a_{m1}x_1 + a_{m2}x_2 + \cdots + a_{mn}x_n = b_m$$

which appears in matrix notation as

$$Ax = b.$$

Here A is a rectangular matrix whose elements are a_{ij} while x and b are column matrices whose elements are x_i and b_i respectively.

Briefly, the most common definitions from matrix algebra are:

Matrix equality	$A = B$	$\rightarrow$	$a_{ij} = b_{ij}$
Zero matrix	$A = 0$	$\rightarrow$	$a_{ij} = 0$
Matrix addition	$A + B = C$	$\rightarrow$	$a_{ij} + b_{ij} = c_{ij}$
Matrix multiplication	$AB = C$	$\rightarrow$	$\Sigma_k a_{ik}b_{kj} = c_{ij}$

Matrix transpose $\tilde{C} = A \;\rightarrow\; (\tilde{C})_{ij} = c_{ji} = a_{ij}$
 (Note that $(\widetilde{AB}) = \tilde{B}\tilde{A}$ and $(\widetilde{ABC}) = \tilde{C}\tilde{B}\tilde{A}$)
Symmetric matrix
 If A is symmetric

$$a_{ij} = a_{ji} \quad \text{or} \quad A = \tilde{A}$$

It has been convenient throughout this book to use partitioned matrices or matrices which have been subdivided into units which are themselves matrices. For example, the equation

$$Ax = b$$

can be partitioned as

$$\begin{bmatrix} A_1 & A_2 \\ A_3 & A_4 \end{bmatrix} \begin{bmatrix} x_1 \\ x_2 \end{bmatrix} = \begin{bmatrix} b_1 \\ b_2 \end{bmatrix},$$

which can then be written

$$A_1 x_1 + A_2 x_2 = b_1 \tag{A.1.1}$$
$$A_3 x_1 + A_4 x_2 = b_2$$

as two *matrix* equations. The use of partitioned matrices is in general straightforward but care must be taken with operations like the transpose. For example, the transpose of Eq. (A.1.1) is

$$\begin{bmatrix} \tilde{x}_1 & \tilde{x}_2 \end{bmatrix} \begin{bmatrix} \tilde{A}_1 & \tilde{A}_3 \\ \tilde{A}_2 & \tilde{A}_4 \end{bmatrix} = \begin{bmatrix} \tilde{b}_1 & \tilde{b}_2 \end{bmatrix},$$

or

$$\tilde{x}\tilde{A} = \tilde{b}.$$

It is sometimes convenient to refer to the elements of a matrix in physical terms. For example, in the equation $Ax = b$, if the elements of x are displacements and the elements of b are forces, both associated with points, an element of A, a_{ij}, can be referred to as the force at point i due to a unit displacement at point j (all other displacements zero). This is easily motivated by setting x to be zero except element $x_i = 1$ and then multiplying Ax to find b which in this case simply turns out to be the ith column of A. For example

$$\begin{bmatrix} a_{11} & a_{12} & a_{13} & a_{14} \\ a_{21} & a_{22} & a_{23} & a_{24} \\ a_{31} & a_{32} & a_{33} & a_{34} \\ a_{41} & a_{42} & a_{43} & a_{44} \end{bmatrix} \begin{bmatrix} 0 \\ 0 \\ 1 \\ 0 \end{bmatrix} = \begin{bmatrix} a_{13} \\ a_{23} \\ a_{33} \\ a_{43} \end{bmatrix}.$$

A.1.2 VECTOR NOTATION

It is assumed that the reader is familar with the treatment of position, force, moment, rotation, displacement, etc., as vectors in a cartesian coordinate system (*see* Fig. A.1.1). While this book uses very little

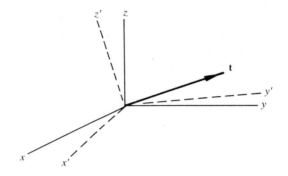

Fig. A.1.1. Cartesian coordinate system.

vector algebra, the transformation from global to local coordinates makes repeated use of the geometrical relationship between the components of vectors in coordinate systems which differ only by a rotation which is stated here for convenience. If **t** is a vector, this relationship is commonly written

$$(\mathbf{t})'_i = \sum_{j=1}^{3} a_{ij}(\mathbf{t})_j \qquad (i = 1, 2, 3) \tag{A.1.2}$$

where

$(\mathbf{t})_i$ — ith component of vector **t** in the unprimed (global) coordinate system

$(\mathbf{t})'_i$ — ith component of vector **t** in the primed (local) coordinate system

a_{ij} — cosine of the angle between the ith axis of the primed coordinate system and the jth axis of the unprimed system

Equation (A.1.2) can be written in a rather obvious manner in matrix notation as

$$t' = At,$$

where

$$t' = \begin{bmatrix} (\mathbf{t})'_x \\ (\mathbf{t})'_y \\ (\mathbf{t})'_z \end{bmatrix}, \qquad t = \begin{bmatrix} (\mathbf{t})_x \\ (\mathbf{t})_y \\ (\mathbf{t})_z \end{bmatrix}, \qquad A = \begin{bmatrix} a_{xx} & a_{xy} & a_{xz} \\ a_{yx} & a_{yy} & a_{yz} \\ a_{zx} & a_{zy} & a_{zz} \end{bmatrix}.$$

(Note that certain liberties are taken here by substituting 1, 2, 3 for x, y, z as subscripts when it is convenient.)

Euler's theorem from kinematics states that the most general displacement of a rigid body with one point fixed can be expressed as a rotation about an axis. While this implies that the transformation from global to local coordinates can always be performed as a rotation about a single axis, it is frequently convenient to decompose this rotation into a compound rotation composed of several rotations about *coordinate axes*. For convenience, the matrix A which describes these rotations is listed here.

Rotation about the x axis of an angle θ,

$$A = \begin{bmatrix} 1 & 0 & 0 \\ 0 & \cos\theta & \sin\theta \\ 0 & -\sin\theta & \cos\theta \end{bmatrix}.$$

Rotation about the y axis of an angle θ,

$$A = \begin{bmatrix} \cos\theta & 0 & -\sin\theta \\ 0 & 1 & 0 \\ \sin\theta & 0 & \cos\theta \end{bmatrix}.$$

Rotation about the z axis of an angle θ,

$$A = \begin{bmatrix} \cos\theta & \sin\theta & 0 \\ -\sin\theta & \cos\theta & 0 \\ 0 & 0 & 1 \end{bmatrix}.$$

The Classical Methods of Structural Analysis

A.2.1 INTRODUCTION

This appendix describes very briefly what might be called the "classical" methods — classical because they were almost universally used prior to the advent of the computer. The reader who is interested in more detail should consult the excellent book of Norris and Wilbur.

The only explanation for the almost complete abandonment of the classical methods appears to be the fact that they are more difficult to program than the methods to which this book is devoted. While the force method described below commonly requires a smaller computational effort (a paramount consideration when calculations must be done by hand), it is not easily programmed to solve an arbitrary structure.

The classical methods are referred to below as the force, displacement, and mixed methods respectively, due to the fact that the unknowns in terms of which the systems of simultaneous linear equations are written are forces, displacements, and some combination of forces and displacements respectively. It should be noted that these linear systems are in general not the same systems which are obtained using the methods described earlier in the book.

A.2.2 THE FORCE METHOD

In the force methods, "releases" such as hinges or "cuts" are introduced into a structure so that it becomes statically determinate. The motivation for this step is the fact that when the structure has been made statically determinate, it is a relatively simple matter to compute displacements in it. But the price paid for the introduction of a release is the development of a discontinuity or "gap" in the structure. In the force method, the forces associated with the releases (the redundant forces) are determined so that the discontinuities are all zero in the final solution.

Consider the case of the plane truss shown in Fig. A.2.1 which is statically indeterminate to the second degree. The system of equations to be solved states that each discontinuity, written as the sum of the discontinuities produced by all possible effects, is zero. For this example these equations are simply

$$\delta_{11}F_1 + \delta_{12}F_2 + \delta_1{}^0 = 0$$
$$\delta_{12}F_1 + \delta_{22}F_2 + \delta_2{}^0 = 0. \tag{A.2.1}$$

In general they can be written as

$$\delta F + \delta^0 = 0 \tag{A.2.2}$$

in which

F_i — unknown redundant force at the ith release

$\delta_i{}^0$ — discontinuity at the ith release due to loads, thermal effects, settlement, etc., with the redundant forces, F_i, zero

δ_{ij} — discontinuity at the ith release due to the effect of a unit value of F_j alone, with all other redundants zero and no loads, thermal effects, etc., present.

A.2.3 THE DISPLACEMENT METHOD

In the displacement method of analysis "constraints" or specified displacements are introduced into a structure so that it becomes "kinematically determinate". This is by no means a unique process but it involves specifying displacements at a sufficient number of points to allow forces to be computed throughout the structure. The price paid for the introduction of a constraint is the development of a fictitious force in the structure. In the displacement method, the displacements associated with the constraints are determined so that the fictitious forces are all zero in the final solution.

Consider the case of the simple plane frame shown in Fig. A.2.2 which is kinematically indeterminate to the third degree when member length changes are not considered. The system of equations to be solved states that each fictitious force, written as the sum of the forces produced by all possible effects, is zero. For this example these equations are just

$$F_{11}\delta_1 + F_{12}\delta_2 + F_{13}\delta_3 + F_1{}^0 = 0,$$
$$F_{21}\delta_1 + F_{22}\delta_2 + F_{23}\delta_3 + F_2{}^0 = 0, \tag{A.2.3}$$
$$F_{31}\delta_1 + F_{32}\delta_2 + F_{33}\delta_3 + F_3{}^0 = 0.$$

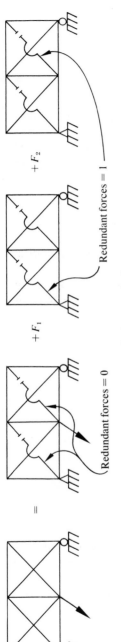

Redundant forces = 0

$+ F_1$

$+ F_2$

Redundant forces = 1

Fig. A.2.1. The force method.

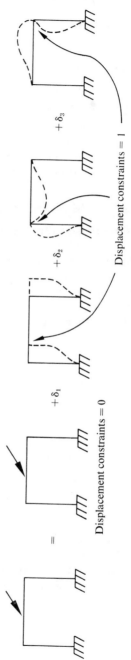

Displacement constraints = 0

$+ \delta_1$

$+ \delta_2$

$+ \delta_3$

Displacement constraints = 1

Fig. A.2.2. The displacement method.

In general they can be written as

$$F\delta + F^0 = 0,\qquad\qquad (A.2.4)$$

in which

δ_i — unknown displacement of the ith constraint

F_i^0 — fictitious forces at the ith constraint due to loads, thermal effects, settlement, etc., with the constraints, δ_i, zero.

F_{ij} — fictitious force at the ith constraint due to the effect of a unit value of δ_j alone, with all other constraints zero and no loads, thermal effects, etc., present.

A.2.4 THE MIXED METHOD

One reason for selecting a particular method over another is that it is easier, which frequently means that it requires less computational effort. For example, in some problems the force method requires the solution of a smaller number of equations than the displacement method and vice versa. There are cases in which it is advantageous to mix the methods and use both forces and displacements as unknowns.

The rather contrived example shown in Fig. A.2.3 is highly statically

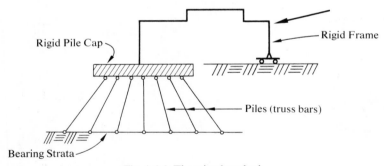

Fig. A.2.3. The mixed method.

and kinematically indeterminate but can be described in terms of the three displacements of the pile cap and the vertical reaction at the roller support, a "mixed" system.

The method has not been commonly used and will be pursued no further here.

Beams and Rods

A.3.1 INTRODUCTION

In the main body of this book it has been assumed that the reader is familiar with some elementary results concerning beams and rods which would ordinarily be included in a first course in mechanics of solids. Those results are briefly reviewed here; for more detail the classical work of Timoshenko can be consulted. The results developed for beams, while not obtained in the most simple manner, are developed in a manner which requires a minimum of background.

A.3.2 RODS

In constructing the primitive stiffness matrix for trusses, the result

$$\Delta_i = \frac{F_i L_i}{A_i E} \qquad (A.3.1)$$

has been used in which

Δ_i — length change of bar i
F_i — force in bar i
L_i — length of bar i
A_i — area of bar i
E — Young's modulus

Equation (A.3.1) may be obtained for a bar under uniaxial stress by first noting that if the bar length change is Δ_i, the bar strain is

$$\epsilon_i = \frac{\Delta_i}{L_i}. \qquad (A.3.2)$$

95

Hooke's law for uniaxial stress simply states that the stress in bar i,

$$\sigma_i = \frac{F_i}{A_i}, \tag{A.3.3}$$

is related to the strain through

$$\sigma_i = E\epsilon_i, \tag{A.3.4}$$

from which Eq. (A.3.1) follows directly.

A.3.3 BEAMS

In this section some well known elementary results from beam theory are developed. The differential equations have been used here as the starting point in an attempt to appeal to as wide a class of readers as possible.

Figure A.3.1 indicates the usual notation of beam theory in which

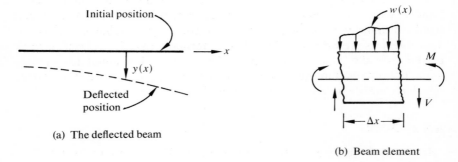

(a) The deflected beam

(b) Beam element

Fig. A.3.1.

a straight, uniform, elastic, plane beam is deformed under lateral load. In the figure,

y — lateral deflection
w — lateral load
M — bending moment
V — shear

In terms of this notation, the beam is described by its equilibrium equations

$$\frac{dV}{dx} + w = 0 \quad \text{and} \quad \frac{dM}{dx} = V \tag{A.3.5}$$

and its constitutive equation

$$M = -EI\frac{d^2y}{dx^2},$$ (A.3.6)

from which follow

$$\frac{d^2M}{dx^2} = -w \quad \text{and} \quad EI\frac{d^4y}{dx^4} = w.$$ (A.3.7)

Figure A.3.2 shows three cases for which results are given.

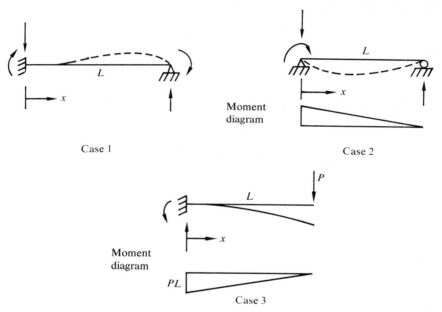

Fig. A.3.2. Three elementary cases.

Case 1

Since $w = 0$, $d^4y/dx^4 = 0$ and $y = a_0 + a_1x + a_2x^2 + a_3x^3$. Applying the boundary conditions,

$$y = 0 \quad \text{at} \quad x = 0$$
$$dy/dx = 0 \quad \text{at} \quad x = 0$$
$$y = 0 \quad \text{at} \quad x = L$$
$$dy/dx = 1 \quad \text{at} \quad x = L$$

it follows that

$$y = \frac{x^2}{L}\left(\frac{x}{L} - 1\right) \quad \text{and} \quad M = -EI\frac{d^2y}{dx^2} = -\frac{EI}{L}\left(6\frac{x}{L} - 2\right).$$

For the stiffness of a plane beam element, the values

$$M_{x=0} = \frac{2EI}{L} \quad \text{and} \quad M_{x=L} = -\frac{4EI}{L}$$

are used.

Case 2

Since Cases 2 and 3 are statically determinate, it is most direct to start with the known moment diagram in these cases and integrate. For a unit applied end moment,

$$M = 1 - \frac{x}{L} = -EI\frac{d^2y}{dx^2}$$

and

$$y = -\frac{1}{EI}\left(\frac{x^2}{2} - \frac{x^3}{6L}\right) + c_1 x + c_2.$$

Applying the boundary conditions, $y = 0$ at $x = 0, L$, results in

$$y = -\frac{L^2}{EI}\left(\frac{x^2}{2L^2} - \frac{x^3}{6L^3} - \frac{x}{3L}\right).$$

At $x = 0$ it follows that

$$\frac{dy}{dx} = \frac{L}{3EI},$$

a result used in Chapter 7.

Case 3

Again starting from the moment diagram and integrating, it follows that

$$M = -PL\left(1 - \frac{x}{L}\right) = -EI\frac{d^2y}{dx^2}$$

and

$$y = \frac{PL}{EI}\left(\frac{x^2}{2} - \frac{x^3}{6L}\right) + c_1 x + c_2.$$

Applying the boundary conditions, $y = 0$ and $dy/dx = 0$ at $x = L$, results in

$$y = \frac{PL^3}{EI}\left(\frac{x^2}{2L^2} - \frac{x^3}{6L^3}\right).$$

At $x = L$

$$y = \frac{PL^3}{3EI},$$

a result used in Chapter 7.

Gaussian Elimination

Programs P.1–P.5 all involve the solution of a system of simultaneous linear algebraic equations. In each case the method used is Gaussian elimination (elimination—backsubstitution). This appendix describes the method briefly.

Gaussian elimination uses two properties of linear algebraic systems:

1. An equation can be multiplied by a constant without changing the values of the unknowns.

2. Linear combinations of equations can be formed without changing the values of the unknowns.

It consists of two steps, elimination and backsubstitution. In the elimination phase the system is reduced to an equivalent upper triangular system; in the backsubstitution phase the unknowns are computed. This is most easily shown by examples but first it should be noted that the scheme used in the computer programs presumes that the system matrix has no zeroes along the diagonal. This is always true for the node method since a zero along a diagonal would indicate an instability in the structure.

Consider the following linear system

$$x_1 + x_2 + x_3 = 1$$
$$2x_1 + x_2 + x_3 = 1$$
$$2x_1 + 3x_2 + x_3 = 1$$

In the elimination phase the terms below the diagonal are removed systematically by forming linear combinations of rows. The terms in the first column below the diagonal are eliminated by subtracting twice the first row from each of the other rows to obtain

$$x_1 + x_2 + x_3 = 1,$$
$$-x_2 - x_3 = -1,$$
$$x_2 - x_3 = -1.$$

The term in the second column below the diagonal can now be eliminated by adding the second and third equations

$$x_1 + x_2 + x_3 = 1,$$
$$-x_2 - x_3 = -1,$$
$$-2x_3 = -2.$$

completing the elimination phase.

In the backsubstitution phase the unknowns are solved for beginning with the last, x_3. The third equation now gives

$$x_3 = 1.$$

Knowing x_3, the second equation can be used to find x_2,

$$x_2 = 1 - x_3 = 0.$$

Finally, the first equation is used to find x_1 given x_2 and x_3,

$$x_1 = 1 - x_2 - x_3 = 0.$$

These steps are carried out for a system of arbitrary size in the programs cited above.

APPENDIX A.5

The Solution of Large Systems

A.5.1 INTRODUCTION

The majority of this text has been concerned with the generation of the system matrix for a structure while little attention has been given to the practical aspects of problem solving. In this appendix, some of the difficulties associated with large systems are discussed together with techniques for the improvement of computational efficiencies.

One of the primary concerns when solving structures is the cost of the solution. This is due to the fact that:

1. Practical problems frequently require the solution of large systems.

2. The cost of computation can be very sensitive to the method of solution used and it is quite common to find order-of-magnitude differences between techniques which produce comparable numerical results.

The success which has been experienced solving large structures is due principally to the fact that as linear systems they are sparse and well-conditioned. When a system is ill-conditioned, difficulties are encountered in applying the usual techniques for its solution; such difficulties are beyond the scope of this book. The remainder of this appendix is devoted to exploiting system sparseness to reduce the cost of solutions.

A.5.2 SPARSE SYSTEMS

A system matrix is defined to be sparse if a large fraction of its elements are zero. Sparseness, a common property shared by many physical systems, is illustrated by the example given in Fig. A.5.1. For this example, the number of movable joints which appear in a single equation

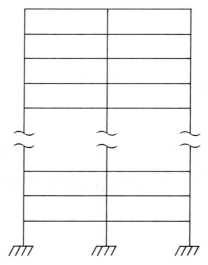

Fig. A.5.1. A building frame.

is at most 5. Clearly, while the size of the system matrix increases linearly with the number of floors, the maximum number of terms in any joint equilibrium equation remains fixed at $5 \times 3 = 15$.

Having established sparseness, the next concern is its utilization. First note that in solving a system of linear simultaneous algebraic equations, a well accepted measure of the computer time required for its solution is obtained by counting the number of necessary multiplications. Sparseness can be utilized by recognizing zero elements and not operating on them during the solution procedure thus reducing the number of multiplications. Sparseness is also utilized by recognizing the zero elements and not storing them in the computer.

A.5.3 AVAILABLE ALTERNATIVES

When solving a system there are several alternatives which present themselves. It is possible to either:

1. Solve the system of simultaneous equations directly.
2. Obtain the inverse system matrix first from which the solution can then be computed.

For each of these schemes it is possible to use

1. Exact methods.
2. Iterative methods.

Briefly, iterative methods such as the Gauss–Siedel procedure appear to converge too slowly to be attractive for use on structural problems where high accuracy is required when forces are computed from displacements by differencing. Iterative methods are not commonly used now to solve structures.

Procedures which involve the computation of the inverse system matrix are also unattractive. First, they are unattractive for reasons of storage requirements since the inverse system matrix for a sparse structure is usually not sparse. Second, from the point of view of counting multiplications, it generally requires more work to solve a system by first computing its inverse system matrix.

By these arguments the available alternatives are reduced to solving a system directly using exact methods.

A.5.4 A GAUSSIAN ELIMINATION SCHEME FOR BAND MATRICES

In this section a Gaussian elimination procedure is described which is "exact" and which does not involve the computation of the inverse system matrix. Furthermore, it is simple and it is compatible with the demands to minimize both computer storage requirements and multiplications required when these two requirements are themselves compatible.

First note that if the rows and columns of the system matrix are properly ordered, the non-zero terms in a sparse system can be concentrated in a band about the diagonal as shown in Fig. A.5.2. The matrix is then referred to as a *band matrix*. Having the terms located within a band facilitates simple computer storage schemes.

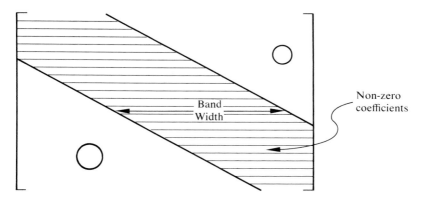

Fig. A.5.2. A band matrix.

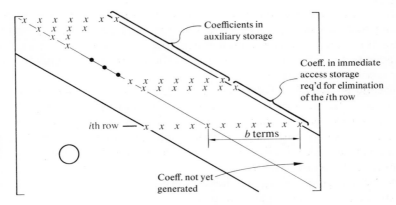

Fig. A.5.3. Elimination for a band matrix.

For a band matrix, there is a simple form of Gaussian elimination which is indicated schematically in Fig. A.5.3. In this elimination-back-substitution scheme, the elimination phase for the ith row consists of generating equation i, then eliminating the terms to the left of the diagonal, then generating equation $i+1$, etc. Note that only b^2 terms are required in immediate access storage for the elimination of the ith row and that once an equation is put into auxiliary storage it is not needed again until the backsubstitution phase.

Computer Program P.6 applies the remarks of this section to large space trusses.

Computer Programs

GENERAL COMMENTS

The following programs are included here to illustrate the methods presented in the main body of this book. The fact that they have been relegated to the location of an appendix should not be taken as an indication of their playing a minor role. Quite to the contrary, they form an integral part of the book: It is one of the strengths of this book that having completed Chapters 1 and 2 the reader is prepared to write or at least understand Programs P.1 and P.2, both of which have quite general capabilities.

These programs are by no means commercial programs. The intent is simplicity; generalization and extensions of these programs make interesting student problems. Like the text, they are best taken in order. For example, while programs one through four have the same structure, the fourth program is most easily understood through an introduction provided by the three programs which precede it.

It should be noted that a computer program does not always follow the formulation in terms of which its programmer may think. The most striking example of this is the fact that neither the matrix K nor the matrix N appear explicitly in the program. Since they are both sparse, their inclusion would constitute a considerable waste of computer storage.

These programs have been used by students at Columbia University on various generations of IBM equipment. In their present form, they run on an IBM 360–91.

Flow Chart

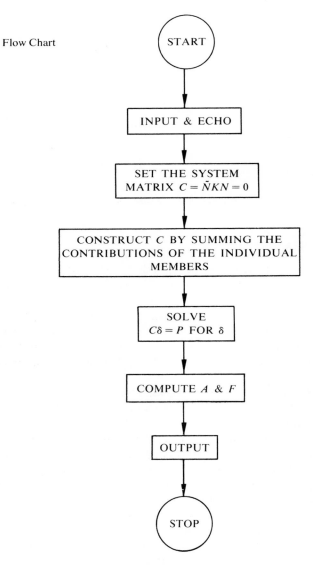

Program P.1. The node method for plane trusses.

P.1 THE NODE METHOD FOR PLANE TRUSSES

This program is a direct application of the material of Chapter 2 to the analysis of an arbitrary plane truss subjected to external joint loads. It requires as input joint coordinates and loads together with member properties; it computes joint displacements and member forces.

Step by Step Explanation of Program P.1

Step 1   Young's modulus. E. is set equal to 30×10^6 psi.

Step 2   Input with echo:
NB — number of truss bars
NN — total number of joints (including supports)
NS — number of supports
R(I) — joint coordinate array (layout similar to the matrix P)
P(I) — joint load matrix P

Step 3   Zero the system matrix $C = \tilde{N}KN$.

Step 4   Construct the matrix C by adding the contributions of the individual members as described under the decomposition procedure. For each member, the quantities
NP(L) — the joint number of the positive end of bar L
NM(L) — the joint number of the negative end of bar L
  S(L) — the area of bar L
are read in and echoed. This step uses the subroutine at the end of the program to generate the unit vector n_L.

Step 5   Solve for the joint displacements δ. A simple Gaussian elimination procedure is used in which the components of δ appear in the array $P(I)$.

Step 6   Compute $\Delta_i = \mathbf{n}_i \cdot (\boldsymbol{\delta}_A - \boldsymbol{\delta}_C)$ and $F_i = (A_i E / L_i) \Delta_i$.

Fortran Program

```
C   PLANE TRUSS
    DIMENSION NP(100),NM(100),S(100)
    DOUBLE PRECISION R(100),P(100),C(100,100),G2,G3,C1,D1,F1,F2
    E=30.E6
100 READ (5,150) NB,NN,NS
    NNN=NN-NS
    READ (5,156) (R(2*I-1),R(2*I),P(2*I-1),P(2*I),I=1,NN)
    WRITE (6,157)
    WRITE (6,158) (I,R(2*I-1),R(2*I),P(2*I-1),P(2*I),I=1,NN)
    N=2*NN
    DO 30 I=1,N
    DO 30 J=1,N
 30 C(I,J)=0.
    N=2*NNN
    WRITE (6,159)
    L1=0
    L=0
201 L=L+1
    READ (5,151) NP(L),NM(L),S(L)
    WRITE (6,160) L,NP(L),NM(L),S(L)
    K=2*NP(L)
    M=2*NM(L)
    GO TO 101
104 Z=-1.
    J=1
 31 GO TO (103,33,34,32,12),J
 32 Z=1.
    K=2*NP(L)
    GO TO 103
 33 Z=1.
    K=2*NM(L)
    GO TO 103
 34 Z=-1.
    M=2*NP(L)
103 C(K-1,M-1) = C(K-1,M-1)+Z*G2*G2*E*S(L)/C1
    C(K-1,M  ) = C(K-1,M  )+Z*G2*G3*E*S(L)/C1
    C(K  ,M-1) = C(K  ,M-1)+Z*G3*G2*E*S(L)/C1
    C(K  ,M  ) = C(K  ,M  )+Z*G3*G3*E*S(L)/C1
    J=J+1
    GO TO 31
 12 IF(L-NB) 201,202,202
202 M=N-1
    DO 17 I=1,M
    L=I+1
    DO 17 J=L,N
    IF (C(J,I))19,17,19
 19 DO 18 K=L,N
 18 C(J,K)=C(J,K)-C(I,K)*C(J,I)/C(I,I)
```

```
       P(J)=P(J)-P(I)                    *C(J,I)/C(I,I)
   17 CONTINUE
       P(N)=P(N)/C(N,N)
       DO 20 I=1,M
       K=N-I
       L=K+1
       DO 21 J=L,N
   21 P(K)=P(K)-P(J)*C(K,J)
       P(K)=P(K)/C(K,K)
   20 CONTINUE
       NS2=2*NS
       DO 204 I=1,NS2
       J=N+I
  204 P(J)=0.
       WRITE (6,161)
       WRITE (6,152)(I,P(2*I-1),P(2*I),I=1,NNN)
       WRITE (6,162)
       L1=1
       I=0
  203 I=I+1
       K=2*NP(I)
       M=2*NM(I)
       F1=P(K-1)-P(M-1)
       F2=P(K)-P(M)
       GO TO 101
  102 D1=F1*G2+F2*G3
       F1=D1*E*S(I)/C1
       F2=F1/S(I)
       WRITE (6,952) I,D1,F1,F2
   22 IF(I-NB)203,100,100
  101 G2=R(K-1)-R(M-1)
       G3=R(K)-R(M)
       C1=DSQRT(G2*G2+G3*G3)
       G2=G2/C1
       G3=G3/C1
       IF(L1) 102,104,102
  150 FORMAT(3(I4,3X))
  151 FORMAT(2I5,8X,F10.6)
  156 FORMAT(    8X,4F11.6)
  152 FORMAT(I10,2D20.8)
  952 FORMAT(I10,3D20.8)
  157    FORMAT(1H1,18X,11HCOORDINATES,32X,5HLOADS/
      114X,1HX,19X,1HY,18X,2HPX,18X,2HPY//)
  158    FORMAT(I4,4D20.8)
  159    FORMAT(1H1,3X,6HMEMBER,5X,5H+ END,5X,5H- END,6X,4HAREA//)
  160    FORMAT(3I10,F20.8)
  161    FORMAT(1H1,13HDISPLACEMENTS/20X,1HX,19X,1HY//)
  162    FORMAT(1H1,3X,6HMEMBER,9X,2HDL,17X,5HFORCE,14X,6HSTRESS//)
       END
```

SOLVE SYSTEM

PRINT δ

COMPUTE F & Δ

SUBROUTINE FOR M.

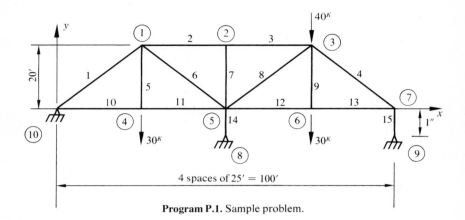

Program P.1. Sample problem.

Comments on the Sample Problem, P.1

This problem is a model of a small continuous truss in which the two roller supports which occur at nodes 5 and 7 have been replaced with very stiff bars. This procedure is consistent with the requirement that any support point be completely fixed (which obviates rollers). It is effective because it is simple but in commercial programs it may be desirable to make special provisions for roller supports rather than contend with the increased problem size which results from the use of these fictitious bars.

Card Input

```
15         10        3
 1        300.          240.              0.             0.
 2        600.          240.              0.             0.
 3        900.          240.              0.         -40000.
 4        300.            0.              0.         -30000.
 5        600.            0.              0.
 6        900.            0.              0.         -30000.
 7       1200.            0.              0.
 8        600.           -1.
 9       1200.           -1.
10          0.            0.
10    1     4.
 1    2     4.
 2    3     4.
 3    7     4.
 1    4     3.
 1    5     3.
 2    5     3.
 3    5     3.
 3    6     3.
10    4     4.
 4    5     4.
 5    6     4.
 6    7     4.
 5    8              1.0E5
 7    9              1.0E5
```

Computer Output

| | COORDINATES | | | LOADS | |
	X	Y		P X	P Y
1	0.30000000D 03	0.24000000D 03		0.0	0.0
2	0.60000000D 03	0.24000000D 03		0.0	0.0
3	0.90000000D 03	0.24000000D 03		0.0	-0.40000000D 05
4	0.30000000D 03	0.0		0.0	-0.30000000D 05
5	0.60000000D 03	0.0		0.0	0.0
6	0.90000000D 03	0.0		0.0	-0.30000000D 05
7	0.12000000D 04	0.0		0.0	0.0
8	0.60000000D 03	-0.10000000D 01		0.0	0.0
9	0.12000000D 04	-0.10000000D 01		0.0	0.0
10	0.0	0.0		0.0	0.0

MEMBER	+ END	- END	AREA
1	10	1	0.40000000E 01
2	1	2	0.40000000E 01
3	2	3	0.40000000E 01
4	3	7	0.40000000E 01
5	1	4	0.30000000E 01
6	1	5	0.30000000E 01
7	2	5	0.30000000E 01
8	3	5	0.30000000E 01
9	3	6	0.30000000E 01
10	10	4	0.40000000E 01
11	4	5	0.40000000E 01
12	5	6	0.40000000E 01
13	6	7	0.40000000E 01
14	5	8	0.10000000E 06
15	7	9	0.10000000E 06

MEMBER	DL	FORCE	STRESS
1	-0.61595654D-01	-0.19239250D 05	-0.48098125D 04
2	0.18633349D-01	0.74533395D 04	0.18633349D 04
3	0.18633349D-01	0.74533395D 04	0.18633349D 04
4	-0.16409565D 00	-0.51254871D 05	-0.12813718D 05
5	0.80000000D-01	0.30000000D 05	0.10000000D 05
6	-0.12287246D 00	-0.28784182D 05	-0.95947273D 04
7	0.16543612D-23	0.62038546D-18	0.20679515D-18
8	-0.25953913D 00	-0.60799803D 05	-0.20266601D 05
9	0.80000000D-01	0.30000000D 05	0.10000000D 05
10	0.37558326D-01	0.15023330D 05	0.37558326D 04
11	0.37558326D-01	0.15023330D 05	0.37558326D 04
12	0.10005833D 00	0.40023330D 05	0.10005833D 05
13	0.10005833D 00	0.40023330D 05	0.10005833D 05
14	-0.18654224D-07	-0.55962672D 05	-0.55962672D 00
15	-0.10672888D-07	-0.32018664D 05	-0.32018664D 00

DISPLACEMENTS

	X	Y
1	0.76794634D-01	-0.19459445D 00
2	0.95427983D-01	-0.18654224D-07
3	0.11406133D 00	-0.46414619D 00
4	0.37558326D-01	-0.27459445D 00
5	0.75116651D-01	-0.18654224D-07
6	0.17517498D 00	-0.54414619D 00
7	0.27523330D 00	-0.10672888D-07

P.2 THE NODE METHOD FOR SPACE TRUSSES

With the exception of increased dimensionality, Program P.2 is identical to Program P.1. The reader is referred to the latter for a flow chart, notes, etc.

Fortran Program

```
C       SPACE TRUSS
        DIMENSION NP(100),NM(100),S(100)
        DOUBLE PRECISION R(100),P(100),C(100,100),G1,G2,G3,C1,D1,F1,F2,F3
        E = 30000000.
100     READ            (5,150)NB,NN,NS
        NNN = NN - NS
        DO 2 I = 1,NN
        K = 3*I
2       READ            (5,156)R(K-2),R(K-1),R(K),P(K-2),P(K-1),P(K)
        WRITE           (6,157)
        DO 3 I = 1,NN
        K = 3*I
3       WRITE           (6,158)I,R(K-2),R(K-1),R(K),P(K-2),P(K-1),P(K)
        N=3*NN
        DO 30 I = 1,N
        DO 30 J = 1,N
30      C(I,J) = 0.
        N = 3*NNN
        L1 = 0
        WRITE           (6,159)
        L=0
201     L=L+1
        READ            (5,151)NP(L),NM(L),S(L)
        WRITE           (6,160)L,NP(L),NM(L),S(L)
        K = 3*NP(L)
        M = 3*NM(L)
        GO TO 101
104     Z = -1.
        J = 1
31      GO TO (103,33,34,32,12),J
32      Z = 1.
        K = 3*NP(L)
        GO TO 103
33      Z = 1.
        K = 3*NM(L)
        GO TO 103
34      Z = -1.
        M = 3*NP(L)
103     C(K-2,M-2) = C(K-2,M-2)+Z*G1*G1*E*S(L)/C1
        C(K-2,M-1) = C(K-2,M-1)+Z*G1*G2*E*S(L)/C1
        C(K-2,M  ) = C(K-2,M  )+Z*G1*G3*E*S(L)/C1
        C(K-1,M-2) = C(K-1,M-2)+Z*G2*G1*E*S(L)/C1
        C(K-1,M-1) = C(K-1,M-1)+Z*G2*G2*E*S(L)/C1
        C(K-1,M  ) = C(K-1,M  )+Z*G2*G3*E*S(L)/C1
        C(K  ,M-2) = C(K  ,M-2)+Z*G3*G1*E*S(L)/C1
        C(K  ,M-1) = C(K  ,M-1)+Z*G3*G2*E*S(L)/C1
        C(K  ,M  ) = C(K  ,M  )+Z*G3*G3*E*S(L)/C1
        J = J + 1
```

INPUT

CONSTRUCT SYSTEM MATRIX

```
      GO TO 31
   12 IF(L-NB) 201,202,202
  202 M = N - 1
      DO 17 I = 1,M
      L = I + 1
      DO 17 J = L,N
      IF(C(J,I)) 19,17,19
   19 DO 18 K = L,N
   18 C(J,K) = C(J,K) - C(I,K)*C(J,I)/C(I,I)
      P(J) = P(J) - P(I)*C(J,I)/C(I,I)
   17 CONTINUE
      P(N) = P(N)/C(N,N)
      DO 20 I = 1,M
      K = N - I
      L = K + 1
      DO 21 J = L,N
   21 P(K) = P(K) - P(J)*C(K,J)
      P(K) = P(K)/C(K,K)
   20 CONTINUE
      NS2=3*NS
      DO 204 I=1,NS2
      J=N+I
  204 P(J)=0.
      WRITE              (6,161)
      WRITE              (6,152)(I,P(3*I-2),P(3*I-1),P(3*I),I=1,NNN)
      WRITE              (6,162)
      L1 = 1
      I = 0
  203 I=I+1
      K = 3*NP(I)
      M = 3*NM(I)
       F1 = P(K-2) - P(M-2)
       F2 = P(K-1) - P(M-1)
       F3 = P(K  ) - P(M  )
      GO TO 101
  102 D1=F1*G1+F2*G2+F3*G3
      F1=D1*E*S(I)/C1
      F2=F1/S(I)
      WRITE(6,1000) I,D1,F1,F2
   22 IF(I-NB) 203,100,100
  101 G1 = R(K-2) - R(M-2)
      G2 = R(K-1) - R(M-1)
      G3 = R(K  ) - R(M  )
      C1=DSQRT(G1*G1+G2*G2+G3*G3)
      G1 = G1/C1
      G2 = G2/C1
      G3 = G3/C1
      IF(L1) 102,104,102
```

SOLVE SYSTEM

PRINT δ

COMPUTE FEA

SUBROUTINE FOR r_i

```
  150    FORMAT (3(I4,3X))
  151    FORMAT (2I5,8X,E10.6)
  156    FORMAT (8X,6F11.6)
  152    FORMAT (I10,3D20.8)
 1000    FORMAT (I10,3D20.8)
  157    FORMAT (1H1,25X,11HCOORDINATES,40X,5HLOADS//
        114X,1HX,19X,1HY,19X,1HZ,18X,2HPX,18X,2HPY,18X,2HPZ//)
  158    FORMAT (I4,6D20.8)
  159    FORMAT (1H1,3X,6HMEMBER,5X,5H+ END,5X,5H- END,6X,4HAREA//)
  160    FORMAT (3I10,E20.8)
  161    FORMAT (1H1,13HDISPLACEMENTS/20X,1HX,19X,1HY,19X,1HZ//)
  162    FORMAT (1H1,3X,6HMEMBER,9X,2HDL,17X,5HFORCE,14X,6HSTRESS//)
         END
```

Sample Problem

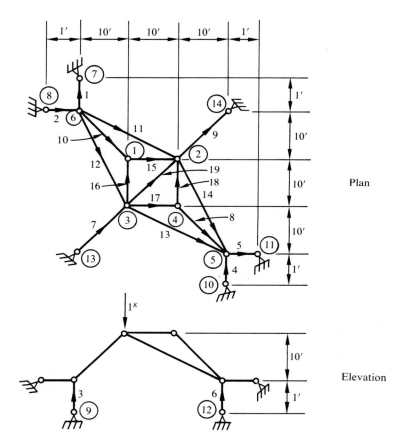

Plan

Elevation

Program P.2. Sample problem.

Card Input

```
/*
//GO.SYSIN      DD    *
   19      14       8
           120.        240.      120.       0.        0.          0.
           240.        240.      120.       0.        0.          0.
           120.        120.      120.       0.        0.      -1000.
           240.        120.      120.       0.        0.          0.
           360.          0.        0.       0.        0.          0.
             0.        360.        0.       0.        0.          0.
             0.        372.        0.       0.        0.          0.
           -12.        360.        0.       0.        0.          0.
             0.        360.      -12.       0.        0.          0.
           360.        -12.        0.       0.        0.          0.
           372.          0.        0.       0.        0.          0.
           360.          0.      -12.       0.        0.          0.
             0.          0.        0.       0.        0.          0.
           360.        360.        0.       0.        0.          0.
    7       6      1200000.
    6       8      1200000.
    6       9      1200000.
    5      10      1200000.
   11       5      1200000.
    5      12      1200000.
    3      13       208.
    5       4       208.
   14       2       208.
    1       6       208.
    2       6       294.
    3       6       294.
    5       3       294.
    5       2       294.
    2       1       120.
    1       3       120.
    4       3       120.
    2       4       120.
    2       3       170.
/*
```

Computer Output

	COORDINATES			LOADS		
	X	Y	Z	PX	PY	PZ
1	0.12000000D 03	0.24000000D 03	0.12000000D 03	0.0	0.0	0.0
2	0.24000000D 03	0.24000000D 03	0.12000000D 03	0.0	0.0	0.0
3	0.12000000D 03	0.12000000D 03	0.12000000D 03	0.0	0.0	0.0
4	0.24000000D 03	0.12000000D 03	0.12000000D 03	0.0	0.0	0.0
5	0.36000000D 03	0.36000000D 03	0.0	0.0	0.0	0.0
6	0.0	0.37200000D 03	0.0	0.0	0.0	0.0
7	0.0	0.36000000D 03	0.0	0.0	0.0	0.0
8	-0.12000000D 02	0.36000000D 03	-0.12000000D 02	0.0	0.0	0.0
9	0.0	-0.12000000D 03	0.0	0.0	0.0	0.0
10	0.36000000D 03	0.0	-0.12000000D 02	0.0	0.0	0.0
11	0.37200000D 03	0.0	0.0	0.0	0.0	0.0
12	0.36000000D 03	0.0	0.0	0.0	0.0	-0.10000000D 04
13	0.0	0.0	0.0	0.0	0.0	0.0
14	0.36000000D 03	0.36000000D 03	0.0	0.0	0.0	0.0

MEMBER	+ END	− END	AREA
1	7	6	0.12000000E 07
2	6	8	0.12000000E 07
3	6	9	0.12000000E 07
4	5	10	0.12000000E 07
5	11	5	0.12000000E 07
6	5	12	0.12000000E 07
7	3	13	0.20800000E 03
8	5	4	0.20800000E 03
9	14	2	0.20800000E 03
10	1	6	0.20800000E 03
11	2	6	0.29400000E 03
12	3	6	0.29400000E 03
13	5	3	0.29400000E 03
14	5	2	0.29400000E 03
15	2	1	0.12000000E 03
16	1	3	0.12000000E 03
17	4	3	0.12000000E 03
18	2	4	0.12000000E 03
19	2	3	0.17000000E 03

DISPLACEMENTS

	X	Y	Z
1	0.40430710D-05	0.70716703D-05	0.30282357D-05
2	0.40430710D-05	0.40430710D-05	0.20229263D-05
3	0.70716703D-05	0.70716703D-05	-0.53515226D-04
4	0.70716703D-05	0.40430710D-05	0.30282357D-05
5	0.19188370D-09	-0.80772584D-10	-0.90885427D-10
6	-0.80772584D-10	0.19188370D-09	-0.90885427D-10

MEMBER	DL	FORCE	STRESS
1	-0.19188370D-09	-0.57565109D 03	-0.47970924D-03
2	-0.80772584D-10	-0.24231775D 03	-0.20193146D-03
3	-0.90885427D-10	-0.27265628D 03	-0.22721357D-03
4	-0.80772584D-10	-0.24231775D 03	-0.20193146D-03
5	-0.19188370D-09	-0.57565109D 03	-0.47970924D-03
6	-0.90885427D-10	-0.27265628D 03	-0.22721357D-03
7	-0.22731368D-04	-0.68244601D 03	-0.32809904D 01
8	-0.21175824D-21	-0.63574511D-14	-0.30564669D-16
9	-0.35005992D-05	-0.10509574D 03	-0.50526798D 00
10	0.84703295D-21	0.25429804D-13	0.12225868D-15
11	0.24766144D-05	0.74313910D 02	0.25276840D 00
12	-0.24734270D-04	-0.74218267D 03	-0.25244309D 01
13	-0.24734270D-04	-0.74218267D 03	-0.25244309D 01
14	0.24766144D-05	0.74313910D 02	0.25276840D 00
15	0.63527471D-21	0.19058241D-13	0.15881868D-15
16	-0.21175824D-20	-0.63527471D-13	-0.52939559D-15
17	0.61409889D-20	0.18422967D-12	0.15352472D-14
18	-0.55057142D-20	-0.16517142D-12	-0.13764285D-14
19	-0.42830861D-05	-0.12871547D 03	-0.75714981D 00

P.3 THE NODE METHOD FOR PLANE FRAMES

This program is a direct application of the material of Chapter 4 to the analysis of an arbitrary plane frame subjected to external joint loads. It requires as input joint loads and member properties; it computes joint displacements and member forces.

Flow Chart

The flow chart for this program is identical with the one given for Program P.1.

Step by Step Explanation of Program P.3

Step 1   Initialize:
  E — Young's modulus to be 30×10^6 psi
  R(I, J) — rotation matrix R_i
  SK(I, J) — member stiffness matrix K_i
  SN(I, J) — matrices N^+ and N^-, depending upon its use

Step 2   Input with echo:
NB — number of members
NN — total number of joints (including supports)
NS — number of supports
P(I) — joint load matrix P

Step 3   Zero the system matrix $C = \tilde{N}KN$.

Step 4   Construct the matrix C by adding the contributions of the individual members as described under the decomposition procedure. For each member, the quantities:
  A(K) — area of member K
  AL(K) — length of member K
  SI(K) — the moment of inertia of member K
  TH(K) — the angle of orientation Φ_K of member K
  LP(K) — the joint number of the positive end of member K
  MI(K) — the joint number of the negative end of member K are read in and echoed. This step uses the subroutines indicated on the program to generate $N_i^{\pm}$, R_i and K_i.

Step 5   Solve for the joint displacements δ. A simple Gaussian elimination procedure is used in which the components of δ appear in the array $P(I)$.

Step 6   Compute $\Delta_i = N_i^+ R_i \delta_A + N_i^- R_i \delta_c$ and $F_i = K_i \Delta_i$.

Fortran Program

```
C       PLANE FRAMES
        DIMENSION A(100),AL(100),SI(100),TH(100),LP(100),MI(100)
        DOUBLE PRECISION P(100),C(100,100),R(3,3),SK(3,3),SN(3,3),AI(3,3)
       1 ,ANG
        R(1,3)=0.
        R(2,3)=0.
        R(3,1)=0.
        R(3,2)=0.
        R(3,3)=1.
        PI=3.14159      /180.
        SK(1,2)=0.
        SN(1,3)=0.
        SN(2,1)=0.
        SN(1,2)=0.
        SK(1,3)=0.
        SK(2,1)=0.
        SN(3,1)=0.
        SK(3,1)=0.
        E=30000000.
      1 LZ1=0
        READ(5,2) NB,NN,NS
        NNS=NN-NS
      2 FORMAT (3(I3,3X))
      3 FORMAT(4(E10.2,2X),2(3X,I3))
        DO 204 I=1,NNS
        J=3*I-2
    204 READ            (5,3)P(J),P(J+1),P(J+2)
        WRITE(6,903)
        WRITE                 (6,902)(I,P(3*I-2),P(3*I-1),P(3*I),I=1,NNS )
    902 FORMAT(I4,3D20.8)
    903 FORMAT(1H1,11HJOINT LOADS /13X,2HPX,18X,2HPY,19X,1HM//)
        WRITE                 (6,901)
        N=3*NN
        DO 904 I=1,N
        DO 904 J=1,N
    904 C(I,J)=0.
        N=3*NNS
        K=0
    926 K=K+1
        READ            (5,3)A(K),AL(K),SI(K),TH(K),LP(K),MI(K)
        WRITE                 (6,900)K, A(K),AL(K),SI(K),TH(K),LP(K),MI(K)
    900 FORMAT(I4,4E20.8,2I10)
    901 FORMAT(18H1MEMBER PROPERTIES /11X,4HAREA,15X,6HLENGTH,16X,1HI,17X,
       1 5HANGLE,14X,5H+ END,5X,5H- END)
        ZK=1.
        ZL=1.
        II=3*LP(K)-3
        JJ=II
```

INITIALIZE ARRAYS

INPUT & ECHO

```
        J=1
   12 GO TO (13,11,9,7,26),J
   11 JJ=3*MI(K)-3
        ZL=-1.
        GO TO 10
    9 II=JJ
        ZK=-1.
        ZL=-1.
        Q=0.
        GO TO 352
    7 JJ=3*LP(K)-3
        ZL=1.
        GO TO 10
   13 ANG=TH(K)*PI
        R(1,1)=DCOS(ANG)
        R(2,2)=R(1,1)
        R(1,2)=DSIN(ANG)
        R(2,1)=-R(1,2)
  214 SK(1,1)=E*A(K)/AL(K)
        SK(2,2)=E*4.*SI(K)/AL(K)
        SK(3,3)=SK(2,2)
        SK(2,3)=SK(2,2)*.5
        SK(3,2)=SK(2,3)
        Q=0.
  352 SN(1,1)=ZK
        SN(2,2)=-ZK/AL(K)
        SN(3,2)=-ZK/AL(K)
        IF(ZK) 15,15,16
   16 SN(2,3)=1.
        SN(3,3)=0.
        GO TO 351
   15 SN(2,3)=0.
        SN(3,3)=1.
  351 IF (LZ1) 71,17,71
   17 IF (Q) 24, 23,24
   23 DO 19 I1=1,3
        DO 19 I2=1,3
   14 AI(I1,I2)=0.
        DO 19 I3=1,3
        DO 19 I4=1,3
   19 AI(I1,I2)=AI(I1,I2)+R(I3,I1)*SN(I4,I3)*SK(I4,I2)
   10 ZK=ZL
        Q=1.
        GO TO 352
   24 DO 27 I1=1,3
        N3=II+I1
        DO 27 I2=1,3
        N4=JJ+I2
```

SUBROUTINES FOR N_i, K_i, R_i

CONSTRUCT SYSTEM MATRIX

```
        DO 27 I3=1,3
        DO 27 I4=1,3
   27 C(N3,N4)=C(N3,N4)+AI(I1,I3)*SN(I3,I4)*R(I4,I2)
        J=J+1
        GO TO 12
   26 IF(K-NB)926,927,927
  927 M=N-1
        DO 91 I=1,M
        L=I+1
        DO 91 J=L,N
        IF (C(J,I)) 93,91,93
   93   DO 92 K=L,N
   92   C(J,K)=C(J,K)-C(I,K)*C(J,I)/C(I,I)
        P(J)=P(J)-P(I)          *C(J,I)/C(I,I)
   91   CONTINUE
        P (N)=P(N)/C(N,N)
        DO 94 I=1,M
        K=N-I
        L=K+1
        DO 95 J=L,N
   95   P(K)=P(K)-P (J)*C(K,J)
   94   P (K)=P(K)/C(K,K)
        WRITE (6,230)
  230 FORMAT(28H1MEMBER DISPLACEMENTS-FORCES/13X,2HDL,15X,3HAL+,16X,
     1 3HAL-,16X,1HT,16X,2HM+,16X,2HM-)
        I=0
  300 I=I+1
        LZ1=1
        ZK=1.
        L1=3*LP(I)-3
        L2=2
        DO 222 J=1,3
  222 AI(L2,J)=0.
        K=I
        L3=LP(I)
        GO TO 13
   71 IF(L3-NNS) 212,212,213
  212 DO 208 J=1,3
  221 DO 208 I1=1,3
        DO 208 L=1,3
        L4=L1+L
  208 AI(L2,J)=AI(L2,J)+SN(J,I1)*R(I1,L)*P(L4)
  213 CONTINUE
        IF(LZ1) 215,210,210
  210 ZK=-1.
        LZ1=-1
        L3=MI(I)
        L1=3*MI(I)-3



        GO TO 352
  215 DO 226 J=1,3
   .    AI(1,J)=0.
        DO 226 K=1,3
  226   AI(1,J)= AI(1,J)+SK(J,K)*AI(2,K)
  207 WRITE            (6,217)I,AI(2,1),AI(2,2),AI(2,3), AI(1,1),
     6 AI(1,2), AI(1,3)
  217 FORMAT (I3,3X,6(D15.8,3X))
        IF(I-NB) 300,301,301
  301 WRITE(6,231)
  231 FORMAT(20H1JOINT DISPLACEMENTS /13X,2HDX,16X,2HDY,16X,2HTH)
        DO 218 I=1,NNS
        J=3*I-3
  218 WRITE            (6,217)I, P(J+1),P(J+2),P(J+3)
        GO TO 1
        END
```

(margin annotations, top to bottom: SOLVE FOR δ ; COMPUTE F&Δ ; OUTPUT)

Sample Problem

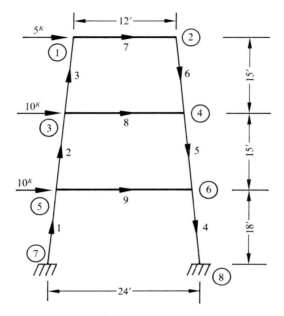

Program P.3. Sample problem.

Card Input

```
9      8      2
    5.00E3      000.00E0      0.00E3
    0.00E3      000.00E0      0.00E3
   10.00E3      000.00E0      0.00E3
    0.00E3      000.00E0      0.00E3
   10.00E3      000.00E0      0.00E3
    0.00E3      000.00E0      0.00E3
    3.00E3      217.68E0      3.00E3      .82.87E0      5      7
    3.00E3      181.40E0      2.00E3      .82.87E0      3      5
    3.00E3      181.40E0      1.00E3      .82.87E0      1      3
    3.00E3      217.68E0      3.00E3     -82.87E0      8      6
    3.00E3      181.40E0      2.00E3     -82.87E0      6      4
    3.00E3      181.40E0      1.00E3     -82.87E0      4      2
    3.00E3      144.00E0      1.00E3      .00.00E0      2      1
    3.00E3      189.00E0      1.00E3      .00.00E0      4      3
    3.00E3      234.00E0      1.00E3      .00.00E0      6      5
        /*
```

Computer Output

JOINT LOADS

	PX	PY	M
1	0.50000000D 04	0.0	0.0
2	0.0	0.0	0.0
3	0.10000000D 05	0.0	0.0
4	0.0	0.0	0.0
5	0.10000000D 05	0.0	0.0
6	0.0	0.0	0.0

MEMBER PROPERTIES

	AREA	LENGTH	I	ANGLE	+ END	- END
1	0.30000000E 04	0.21767999E 03	0.30000000E 04	0.82869995E 02	5	7
2	0.30000000E 04	0.18139999E 03	0.20000000E 04	0.82869995E 02	3	5
3	0.30000000E 04	0.21767999E 03	0.10000000E 04	0.82869995E 02	1	3
4	0.30000000E 04	0.18139999E 03	0.30000000E 04	-0.82869995E 02	8	6
5	0.30000000E 04	0.18139999E 03	0.20000000E 04	-0.82869995E 02	6	4
6	0.30000000E 04	0.14400000E 03	0.10000000E 04	-0.82869995E 02	4	2
7	0.30000000E 04	0.18900000E 03	0.10000000E 04	0.0	2	1
8	0.30000000E 04	0.23400000E 03	0.10000000E 04	0.0	4	3
9	0.30000000E 04		0.10000000E 04	0.0	6	5

MEMBER DISPLACEMENTS-FORCES

	DL	AL+	AL-	T	M+	M-
1	0.53902478D-04	-0.80600216D-04	0.94142731D-03	0.22286020D 05	0.64517108D 06	0.14902876D 07
2	0.27376710D-04	0.60572472D-03	-0.70163308D-04	0.13582710D 05	0.75498520D 06	0.30787081D 06
3	0.91694587D-05	0.57338417D-03	-0.21683560D-03	0.45493449D 04	0.30758509D 06	0.46211544D 05
4	-0.53902219D-04	0.94136713D-03	-0.80615084D-04	-0.22285913D 05	0.14901758D 07	0.64509673D 06
5	-0.27376809D-04	-0.70104119D-04	0.60570117D-03	-0.13582759D 05	0.30793354D 06	0.75499320D 06
6	-0.91694767D-05	-0.21683691D-03	0.57339853D-03	-0.45493538D 04	0.46215425D 05	0.30759415D 06
7	-0.40000579D-05	-0.24608267D-03	-0.24606092D-03	-0.25000360D 04	-0.30759415D 06	-0.30758509D 06
8	-0.10500337D-04	-0.84128212D-03	-0.84124470D-03	-0.50001605D 04	-0.80120863D 06	-0.80119675D 06
9	-0.12998376D-04	-0.12389249D-02	-0.12389703D-02	-0.49993726D 04	-0.95303027D 06	-0.95304188D 06

JOINT DISPLACEMENTS

	DX	DY	TH
1	0.39796373D 00	-0.49690418D-01	0.44408026D-03
2	0.39795973D 00	0.49689918D-01	0.44405851D-03
3	0.37468825D 00	-0.46788112D-01	-0.34613950D-03
4	0.37467775D 00	-0.46786798D-01	-0.34617693D-03
5	0.20335190D 00	-0.25383113D-01	-0.10202275D-02
6	0.20333890D 00	0.25381488D-01	-0.10219822D-02

P.4 THE NODE METHOD FOR SPACE FRAMES

This program is a direct application of the material of Chapter 5 to the analysis of an arbitrary space frame subjected to external joint loads. It requires as input joint loads and member properties; it computes joint displacements and member forces.

Flow Chart

For a flow chart and general notes see Program P.1. Programs P.1–P.4 have the same basic structure and only differ in detail.

Step by Step Explanation of Program P.4

Step 1   Young's Modulus. E. is set equal to 30×10^6 psi; Shear modulus. G. is set equal to 12×10^6 psi; PI is set equal to π.

Step 2   Input with echo:
NB — number of bars
NN — total number of nodes (including supports)
NS — number of supports
P(I) — joint load matrix P

Step 3   Initialize arrays:
   C(I, J) — the system matrix $C = \tilde{N}KN$ is set equal to zero
   R(I, J) — rotation matrix R_i

$$R - \begin{bmatrix} RI & 0 \\ \hline 0 & RI \end{bmatrix}$$

   RI — a three dimensional rotation matrix is constructed later in the program as the product of rotations about axes.
   AK(I, J) — stiffness matrix K_i
AN(I, J, K) — $N_i^{\pm}$ depending upon whether I is one or two

Step 4   Construct the matrix C by adding the contributions of the individual members as described under the decomposition procedure. For each member, the quantities
   AI(I) — area of member I
   AL(I) — length of member I
   AJ(I) — torsional moment of inertia of member I
   AI2(I) — moment of inertia about the y^i axis of member I
   AI3(I) — moment of inertia about the z^i axis of member I
   NP(I) — joint number of the positive end of member I

MI(I) — joint number of the negative end of member I

NT(I, J) — axis of rotation for the Jth rotation of member I

TH(I, J) — value of the Jth rotation about axis $NT(I, J)$ of member I

Note that in this program the orientation of each member is described by prescribing the rotation of the global coordinate system into the local coordinate system through a sequence of rotations about axes. First there is a rotation of $TH(I, 1)$ about the axis $NT(I, 1)$, then a rotation of $TH(I, 2)$ about axis $NT(I, 2)$, etc. The axis designation used is:

axis	designation
x axis	1
y axis	2
z axis	3

Step 5   Solve for the joint displacements δ. A simple Gaussian elimination procedure is used in which the components of δ appear in the array $P(I)$.

Step 6   Compute $\Delta_i = N_i{}^+ R_i \delta_A + N_i{}^- R_i \delta_c$ and $F_i = K_i \Delta_i$.

Fortran Program

```
C       SPACE FRAMES
        DIMENSION A1(100),AL(100),NP(100),MI(100),TH(3,100),NT(3,100),
       1 AJ(100),AI2(100),AI3(100)
        DOUBLE PRECISION P(150),C(150,150),R1(3,3,3),AN(2,6,6),AK(6,6),
       1 AB(6,6),R(6,6),A(6,6),AI(6,6),ANG
        E=30000000.
        PI=3.14159
        G=12000000.
 1000   READ            (5,2)NB,NN,NS
    2   FORMAT (3(I3,3X))
        NNS=NN-NS
        WRITE(6,65)
   65   FORMAT(22H1APPLIED LOADS-MOMENTS /6H JOINT,9X,2HPX,17X,2HPY,17X,
       1 2HPZ,17X,2HMX,17X,2HMY,17X,2HMZ)
  114   FORMAT(I6,6D19.8)
        DO72 J = 1,NNS
        I=6*J-6
   56   FORMAT(I5,2X,6D10.4)
        READ            (5,56)K,P(I+1),P(I+2),P(I+3),P(I+4),P(I+5),P(I+6)
   72   WRITE           (6,114)J,P(I+1),P(I+2),P(I+3),P(I+4),P(I+5),P(I+6)
        N=6*(NNS+1)
        DO 22 I=1,N
        DO 22 J=1,N
   22   C(I,J)=0.
        N=6*NNS
        DO 616 I=1,3
        DO 617 J=1,3
        DO 617 K=1,3
  617   R1(I,J,K)=0.
  616   R1(I,I,I)=1.
        DO 610  I=1,6
        DO 610 J=1,6
        DO 620 K=1,2
  620   AN(K,I,J)=0.
        AK(I,J)=0.
  610   R(I,J)=0.
        AN(1,1,1)=1.
        AN(1,2,4)=1.
        AN(1,3,5)=1.
        AN(1,4,6)=1.
        AN(2,1,1)=-1.
        AN(2,2,4)=-1.
        AN(2,5,5)= 1.
        AN(2,6,6)= 1.
        WRITE           (6,15)
   15   FORMAT(18H1MEMBER PROPERTIES /12X,4HAREA,6X,6HLENGTH,10X,2HIX,
       110X,2HIY,10X,2HIZ,3X,5H+ END,3X,5H- END,3(3X,9HAXIS-DEG.,1X))
   13   JJ2=1
```

(right margin annotations, bottom to top): INITIALIZE ARRAYS / INPUT & ECHO

```
      JZ=0
  64  JZ=JZ+1
      READ            (5,51)A1(JZ),AL(JZ),AJ(JZ),AI2(JZ),AI3(JZ),NP(JZ),
    $ MI(JZ),NT(1,JZ),TH(1,JZ),NT(2,JZ),TH(2,JZ),NT(3,JZ),TH(3,JZ)
      WRITE           (6,104)JZ,A1(JZ),AL(JZ),AJ(JZ),AI2(JZ),AI3(JZ),NP(JZ),
    $), MI(JZ),NT(1,JZ),TH(1,JZ),NT(2,JZ),TH(2,JZ),NT(3,JZ),TH(3,JZ)
 104  FORMAT(I4,5E12.4,2I8,3(I5,F8.2))
      GO TO 602
 623  NZ=1
      NZ1=2
      NF=6*NP(JZ)-6
      I1=6*MI(JZ)-6
      IF(NP(JZ)-NNS)26,26,36
  36  NF=N
  26  IF(MI(JZ)-NNS)633,633,33
  33  I1=N
 633  DO 10 I=1,6
      DO 10 J=1,6
      A(I,J)=0.
      AB(I,J)=0.
      DO 10 J1=1,6
      AB(I,J)=AB(I,J)+AN(NZ1,I,J1)*R(J1,J)
  10  A(I,J)=A(I,J)+AN(NZ,I,J1)*R(J1,J)
      DO 720 I=1,6
      DO 720 J=1,6
      AI(I,J)=0.
      DO 720 J1=1,6
 720  AI(I,J)=AI(I,J)+A(J1,I)*AK(J1,J)
      DO 721 I=1,6
      DO 721 J=1,6
      NFI=NF+I
      NFJ=NF+J
      I1J=I1+J
      DO 721 J1=1,6
      C(NFI,I1J)=C(NFI,I1J)+AI(I,J1)*AB(J1,J)
 721  C(NFI,NFJ)=C(NFI,NFJ)+AI(I,J1)*A(J1,J)
  42  IF(NZ-2) 41,4,41
  41  NZ=2
      NZ1=1
      I11=I1
      I1=NF
      NF=I11
      GO TO 633
   4  IF(JZ-NB)64,301,301
 301  M=N-1
      DO 91 I=1,M
      L=I+1
      DO 91 J=L,N
```

CONSTRUCT SYSTEM MATRIX

```
      IF (C(J,I)) 93,91,93
93    DO 92 K=L,N
92    C(J,K)=C(J,K)-C(I,K)*C(J,I)/C(I,I)
      P(J)=P(J)-P(I)            *C(J,I)/C(I,I)
91    CONTINUE
      P (N)=P(N)/C(N,N)
      DO 94 I=1,M
      K=N-I
      L=K+1
      DO 95 J=L,N
95    P(K)=P(K)-P (J)*C(K,J)
94    P (K)=P(K)/C(K,K)
      WRITE            (6,54)
      DO 35IY=1,NNS
      IZ=6*IY-6
35    WRITE            (6,52)IY,P(IZ+1),P(IZ+2),P(IZ+3)
     1,P(IZ+4),P(IZ+5),P(IZ+6)
      WRITE            (6,55)
      DO630 I=1,6
      J=6*NNS+I
630   P(J)=0.
      JJ2=2
      JZ=0
399   JZ=JZ+1
732   I=JZ
      L1=NP(I)
      IF(L1-NNS)801,801,802
802   L1=NNS+1
801   L2=MI(I)
      IF(L2-NNS)803,803,804
804   L2=NNS+1
803   CONTINUE
      GO TO 602
631   DO 47 J=1,6
      A(J,1)=0.
      A(J,2)=0.
      DO 47 J1=1,6
      J11=6*L1-6+J1
      J12=6*L2-6+J1
      A(J,1)=A(J,1)+R(J,J1)*P(J11)
47    A(J,2)=A(J,2)+R(J,J1)*P(J12)
      DO 722 J=1,6
      A(J,3)=0.
      DO 722 J1=1,6
722   A(J,3)=A(J,3)+AN(1,J,J1)*A(J1,1)+AN(2,J,J1)*A(J1,2)
      DO 723 J=1,6
      A(J,1)=0.
      DO 723 J1=1,6
```

SOLVE SYSTEM

COMPUTE F & D

```
 723 A(J,1)=A(J,1)+AK(J,J1)*A(J1,3)
  44 WRITE            (6,52)JZ,(A(J,1),J=1,6),NP(JZ),MI(JZ)
     IF(JZ-NB)399,1000,1000
 602 JP = JZ
     DO 604 I=1,3
     DO 603 J=1,3
 603 R(J,I)=0.
 604 R(I,I)=1.
     I=0
 302 I=I+1
     IF(TH(I,JP))606,605,606
 606 L=NT(I,JP)
     GO TO 612
 618 DO 607 J=1,3
     DO 607 JA=1,3
     A(J,JA)=0.
     DO 607 JB=1,3
 607 A(J,JA)=A(J,JA)+R1(L,J,JB)*R(JB,JA)
     DO 608 K=1,3
     DO 608 J=1,3
 608 R(K,J)=A(K,J)
 605 IF(I-3) 302,303,303
 303 DO 609 I=1,3
     DO 609 J=1,3
 609 R(I+3,J+3)=R(I,J)
 611 ANG=1./AL(JP)
     DO 621 I=1,2
     AN(I,3,3)=ANG
     AN(I,4,2)=-ANG
     AN(I,5,3)= ANG
     AN(I,6,2)=-ANG
 621 ANG=-ANG
 622 AK(1,1)=A1(JP)*E/AL(JP)
     AK(2,2)=AJ(JP)*G/AL(JP)
     AK(3,3)=4.*E*AI2(JP)/AL(JP)
     AK(5,5)=AK(3,3)
     AK(3,5)=.5*AK(5,5)
     AK(5,3)=AK(3,5)
     AK(4,4)=4.*E*AI3(JP)/AL(JP)
     AK(6,6)=AK(4,4)
     AK(4,6)=.5*AK(6,6)
     AK(6,4)=AK(4,6)
     GO TO (623,631),JJ2
 612 ANG=TH(I,JP)*PI/180.
     IF(L-2)613,614,615
 613 R1(1,2,2)=DCOS(ANG)
     R1(1,2,3)=DSIN(ANG)
     R1(1,3,3)=R1(1,2,2)
```

SUBROUTINES FOR R_i, K_i, & $N_i^{\pm}$

```
      R1(1,3,2)=-R1(1,2,3)
      GO TO 618
  614 R1(2,1,1)=DCOS(ANG)
      R1(2,1,3)=-DSIN(ANG)
      R1(2,3,1)=-R1(2,1,3)
      R1(2,3,3)=R1(2,1,1)
      GO TO 618
  615 R1(3,1,1)=DCOS(ANG)
      R1(3,1,2)=DSIN(ANG)
      R1(3,2,1)=-R1(3,1,2)
      R1(3,2,2)=R1(3,1,1)
      GO TO 618
   50 FORMAT(4(I4,3X))
   51  FORMAT(5E8.4,2I4,3(I4,F6.2))
   52 FORMAT(I4,2X,6(E16.8,1X),2(2X,I4))
   54 FORMAT(1H1,19HJOINT DISPLACEMENTS/
    1         6H JOINT,9X,1HX,17X,1HY,17X,1HZ,14X,4HTH X,14X,4HTH Y,
    1 14X,4HTH Z)
   55 FORMAT(1H1,6HMEMBER,5X,6HTHRUST,11X,6HTORQUE,12X,5HMY(+),12X,
    1 5HMZ(+),12X,5HMY(-),12X,5HMZ(-),7X,6HJOINTS)
   57 FORMAT(I6,E18.8,E18.8,I8,I8)
      END
```

Sample Problem

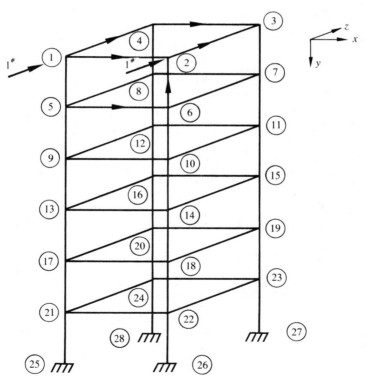

Program P.4. Sample problem.

131

Card Input

```
/ *
//GO.SYSIN DD *
48    24
1       0.0E0       0.0E0       1.0E0       0.0E0       0.0E0       0.0E0
1       0.0E0       0.0E0       1.0E0       0.0E0       0.0E0       0.0E0
1       0.0E0       0.0E0       0.0E0       0.0E0       0.0E0       0.0E0
1       0.0E0       0.0E0       0.0E0       0.0E0       0.0E0       0.0E0
1       0.0E0       0.0E0       0.0E0       0.0E0       0.0E0       0.0E0
1       0.0E0       0.0E0       0.0E0       0.0E0       0.0E0       0.0E0
1       0.0E0       0.0E0       0.0E0       0.0E0       0.0E0       0.0E0
1       0.0E0       0.0E0       0.0E0       0.0E0       0.0E0       0.0E0
1       0.0E0       0.0E0       0.0E0       0.0E0       0.0E0       0.0E0
1       0.0E0       0.0E0       0.0E0       0.0E0       0.0E0       0.0E0
1       0.0E0       0.0E0       0.0E0       0.0E0       0.0E0       0.0E0
1       0.0E0       0.0E0       0.0E0       0.0E0       0.0E0       0.0E0
1       0.0E0       0.0E0       0.0E0       0.0E0       0.0E0       0.0E0
1       0.0E0       0.0E0       0.0E0       0.0E0       0.0E0       0.0E0
1       0.0E0       0.0E0       0.0E0       0.0E0       0.0E0       0.0E0
1       0.0E0       0.0E0       0.0E0       0.0E0       0.0E0       0.0E0
1       0.0E0       0.0E0       0.0E0       0.0E0       0.0E0       0.0E0
1       0.0E0       0.0E0       0.0E0       0.0E0       0.0E0       0.0E0
1       0.0E0       0.0E0       0.0E0       0.0E0       0.0E0       0.0E0
1       0.0E0       0.0E0       0.0E0       0.0E0       0.0E0       0.0E0
1       0.0E0       0.0E0       0.0E0       0.0E0       0.0E0       0.0E0
1       0.0E0       0.0E0       0.0E0       0.0E0       0.0E0       0.0E0
1       0.0E0       0.0E0       0.0E0       0.0E0       0.0E0       0.0E0
1       0.0E0       0.0E0       0.0E0       0.0E0       0.0E0       0.0E0

1.0E0   1.0E2   2.0E0   1.0E0   1.0E0    2    1   1   0.00   1   0.00   1   0.00
1.0E0   1.0E2   2.0E0   1.0E0   1.0E0    3    4   1   0.00   1   0.00   1   0.00
1.0E0   1.0E2   2.0E0   1.0E0   1.0E0    6    5   1   0.00   1   0.00   1   0.00
1.0E0   1.0E2   2.0E0   1.0E0   1.0E0    7    8   1   0.00   1   0.00   1   0.00
1.0E0   1.0E2   2.0E0   1.0E0   1.0E0   10    9   1   0.00   1   0.00   1   0.00
1.0E0   1.0E2   2.0E0   1.0E0   1.0E0   11   12   1   0.00   1   0.00   1   0.00
1.0E0   1.0E2   2.0E0   1.0E0   1.0E0   14   13   1   0.00   1   0.00   1   0.00
1.0E0   1.0E2   2.0E0   1.0E0   1.0E0   15   16   1   0.00   1   0.00   1   0.00
1.0E0   1.0E2   2.0E0   1.0E0   1.0E0   18   17   1   0.00   1   0.00   1   0.00
1.0E0   1.0E2   2.0E0   1.0E0   1.0E0   19   20   1   0.00   1   0.00   1   0.00
1.0E0   1.0E2   2.0E0   1.0E0   1.0E0   22   21   1   0.00   1   0.00   1   0.00
1.0E0   1.0E2   2.0E0   1.0E0   1.0E0   23   24   1   0.00   1   0.00   1   0.00
1.0E0   1.0E2   2.0E0   1.0E0   1.0E0    4    1   2-90.00   1   0.00   1   0.00
1.0E0   1.0E2   2.0E0   1.0E0   1.0E0    3    2   2-90.00   1   0.00   1   0.00
1.0E0   1.0E2   2.0E0   1.0E0   1.0.0    8    5   2-90.00   1   0.00   1   0.00
1.0E0   1.0E2   2.0E0   1.0E0   1.0E0    7    6   2-90.00   1   0.00   1   0.00
1.0E0   1.0E2   2.0E0   1.0E0   1.0E0   12    9   2-90.00   1   0.00   1   0.00
1.0E0   1.0E2   2.0E0   1.0E0   1.0E0   11   10   2-90.00   1   0.00   1   0.00
1.0E0   1.0E2   2.0E0   1.0E0   1.0E0   16   13   2-90.00   1   0.00   1   0.00
1.0E0   1.0E2   2.0E0   1.0E0   1.0E0   15   14   2-90.00   1   0.00   1   0.00
1.0E0   1.0E2   2.0E0   1.0E0   1.0E0   20   17   2-90.00   1   0.00   1   0.00
1.0E0   1.0E2   2.0E0   1.0E0   1.0E0   19   18   2-90.00   1   0.00   1   0.00
1.0E0   1.0E2   2.0E0   1.0E0   1.0E0   24   21   2-90.00   1   0.00   1   0.00
1.0E0   1.0E2   2.0E0   1.0E0   1.0E0   23   22   2-90.00   1   0.00   1   0.00
1.0E0   1.0E2   2.0E0   1.0E0   1.0E0    1    5   3-90.00   1   0.00   1   0.00
1.0E0   1.0E2   2.0E0   1.0E0   1.0E0    5    9   3-90.00   1   0.00   1   0.00
1.0E0   1.0E2   2.0E0   1.0E0   1.0E0    9   13   3-90.00   1   0.00   1   0.00
1.0E0   1.0E2   2.0E0   1.0E0   1.0E0   13   17   3-90.00   1   0.00   1   0.00
1.0E0   1.0E2   2.0E0   1.0E0   1.0E0   17   21   3-90.00   1   0.00   1   0.00
1.0E0   1.0E2   2.0E0   1.0E0   1.0E0   21   25   3-90.00   1   0.00   1   0.00
1.0E0   1.0E2   2.0E0   1.0E0   1.0E0    2    6   3-90.00   1   0.00   1   0.00
1.0E0   1.0E2   2.0E0   1.0E0   1.0E0    6   10   3-90.00   1   0.00   1   0.00
1.0E0   1.0E2   2.0E0   1.0E0   1.0E0   10   14   3-90.00   1   0.00   1   0.00
1.0E0   1.0E2   2.0E0   1.0E0   1.0E0   14   18   3-90.00   1   0.00   1   0.00
1.0E0   1.0E2   2.0E0   1.0E0   1.0E0   18   22   3-90.00   1   0.00   1   0.00
1.0E0   1.0E2   2.0E0   1.0E0   1.0E0   22   26   3-90.00   1   0.00   1   0.00
1.0E0   1.0E2   2.0E0   1.0E0   1.0E0    4    8   3-90.00   1   0.00   1   0.00
1.0E0   1.0E2   2.0E0   1.0E0   1.0E0    8   12   3-90.00   1   0.00   1   0.00
1.0E0   1.0E2   2.0E0   1.0E0   1.0E0   12   16   3-90.00   1   0.00   1   0.00
1.0E0   1.0E2   2.0E0   1.0E0   1.0E0   16   20   3-90.00   1   0.00   1   0.00
1.0E0   1.0E2   2.0E0   1.0E0   1.0E0   20   24   3-90.00   1   0.00   1   0.00
1.0E0   1.0E2   2.0E0   1.0E0   1.0E0   24   28   3-90.00   1   0.00   1   0.00
1.0E0   1.0E2   2.0E0   1.0E0   1.0E0    3    7   3-90.00   1   0.00   1   0.00
1.0E0   1.0E2   2.0E0   1.0E0   1.0E0    7   11   3-90.00   1   0.00   1   0.00
1.0E0   1.0E2   2.0E0   1.0E0   1.0E0   11   15   3-90.00   1   0.00   1   0.00
1.0E0   1.0E2   2.0E0   1.0E0   1.0E0   15   19   3-90.00   1   0.00   1   0.00
1.0E0   1.0E2   2.0E0   1.0E0   1.0E0   19   23   3-90.00   1   0.00   1   0.00
1.0E0   1.0E2   2.0E0   1.0E0   1.0E0   23   27   3-90.00   1   0.00   1   0.00
/ *
```

Computer Output

APPLIED LOADS-MOMENTS

JOINT	PX	PY	PZ	MX	MY	MZ
1	0.0	0.0	0.10000000D 01	0.0	0.0	0.0
2	0.0	0.0	0.10000000D 01	0.0	0.0	0.0
3	0.0	0.0	0.0	0.0	0.0	0.0
4	0.0	0.0	0.0	0.0	0.0	0.0
5	0.0	0.0	0.0	0.0	0.0	0.0
6	0.0	0.0	0.0	0.0	0.0	0.0
7	0.0	0.0	0.0	0.0	0.0	0.0
8	0.0	0.0	0.0	0.0	0.0	0.0
9	0.0	0.0	0.0	0.0	0.0	0.0
10	0.0	0.0	0.0	0.0	0.0	0.0
11	0.0	0.0	0.0	0.0	0.0	0.0
12	0.0	0.0	0.0	0.0	0.0	0.0
13	0.0	0.0	0.0	0.0	0.0	0.0
14	0.0	0.0	0.0	0.0	0.0	0.0
15	0.0	0.0	0.0	0.0	0.0	0.0
16	0.0	0.0	0.0	0.0	0.0	0.0
17	0.0	0.0	0.0	0.0	0.0	0.0
18	0.0	0.0	0.0	0.0	0.0	0.0
19	0.0	0.0	0.0	0.0	0.0	0.0
20	0.0	0.0	0.0	0.0	0.0	0.0
21	0.0	0.0	0.0	0.0	0.0	0.0
22	0.0	0.0	0.0	0.0	0.0	0.0
23	0.0	0.0	0.0	0.0	0.0	0.0
24	0.0	0.0	0.0	0.0	0.0	0.0

MEMBER PROPERTIES

	AREA	LENGTH	IX	IY	IZ	+ END	- END	AXIS-DEG.	AXIS-DEG.	AXIS-DEG.
1	0.1000E 01	0.1000E 03	0.2000E 01	0.1000E 01	0.1000E 01	2	1	1 0.0	1 0.0	1 0.0
2	0.1000E 01	0.1000E 03	0.2000E 01	0.1000E 01	0.1000E 01	3	4	1 0.0	1 0.0	1 0.0
3	0.1000E 01	0.1000E 03	0.2000E 01	0.1000E 01	0.1000E 01	6	5	1 0.0	1 0.0	1 0.0
4	0.1000E 01	0.1000E 03	0.2000E 01	0.1000E 01	0.1000E 01	7	8	1 0.0	1 0.0	1 0.0
5	0.1000E 01	0.1000E 03	0.2000E 01	0.1000E 01	0.1000E 01	10	9	1 0.0	1 0.0	1 0.0
6	0.1000E 01	0.1000E 03	0.2000E 01	0.1000E 01	0.1000E 01	11	12	1 0.0	1 0.0	1 0.0
7	0.1000E 01	0.1000E 03	0.2000E 01	0.1000E 01	0.1000E 01	14	13	1 0.0	1 0.0	1 0.0
8	0.1000E 01	0.1000E 03	0.2000E 01	0.1000E 01	0.1000E 01	15	16	1 0.0	1 0.0	1 0.0
9	0.1000E 01	0.1000E 03	0.2000E 01	0.1000E 01	0.1000E 01	18	17	1 0.0	1 0.0	1 0.0
10	0.1000E 01	0.1000E 03	0.2000E 01	0.1000E 01	0.1000E 01	19	20	1 0.0	1 0.0	1 0.0
11	0.1000E 01	0.1000E 03	0.2000E 01	0.1000E 01	0.1000E 01	22	21	1 0.0	1 0.0	1 0.0
12	0.1000E 01	0.1000E 03	0.2000E 01	0.1000E 01	0.1000E 01	23	24	1 0.0	1 0.0	1 0.0
13	0.1000E 01	0.1000E 03	0.2000E 01	0.1000E 01	0.1000E 01	4	1	2 -90.00	1 0.0	1 0.0
14	0.1000E 01	0.1000E 03	0.2000E 01	0.1000E 01	0.1000E 01	3	2	2 -90.00	1 0.0	1 0.0
15	0.1000E 01	0.1000E 03	0.2000E 01	0.1000E 01	0.1000E 01	8	5	2 -90.00	1 0.0	1 0.0
16	0.1000E 01	0.1000E 03	0.2000E 01	0.1000E 01	0.1000E 01	7	6	2 -90.00	1 0.0	1 0.0
17	0.1000E 01	0.1000E 03	0.2000E 01	0.1000E 01	0.1000E 01	12	9	2 -90.00	1 0.0	1 0.0
18	0.1000E 01	0.1000E 03	0.2000E 01	0.1000E 01	0.1000E 01	11	10	2 -90.00	1 0.0	1 0.0
19	0.1000E 01	0.1000E 03	0.2000E 01	0.1000E 01	0.1000E 01	16	13	2 -90.00	1 0.0	1 0.0
20	0.1000E 01	0.1000E 03	0.2000E 01	0.1000E 01	0.1000E 01	15	14	2 -90.00	1 0.0	1 0.0
21	0.1000E 01	0.1000E 03	0.2000E 01	0.1000E 01	0.1000E 01	20	17	2 -90.00	1 0.0	1 0.0
22	0.1000E 01	0.1000E 03	0.2000E 01	0.1000E 01	0.1000E 01	19	18	2 -90.00	1 0.0	1 0.0
23	0.1000E 01	0.1000E 03	0.2000E 01	0.1000E 01	0.1000E 01	24	21	2 -90.00	1 0.0	1 0.0
24	0.1000E 01	0.1000E 03	0.2000E 01	0.1000E 01	0.1000E 01	23	22	2 -90.00	1 0.0	1 0.0
25	0.1000E 01	0.1000E 03	0.2000E 01	0.1000E 01	0.1000E 01	1	5	3 -90.00	1 0.0	1 0.0
26	0.1000E 01	0.1000E 03	0.2000E 01	0.1000E 01	0.1000E 01	5	9	3 -90.00	1 0.0	1 0.0
27	0.1000E 01	0.1000E 03	0.2000E 01	0.1000E 01	0.1000E 01	9	13	3 -90.00	1 0.0	1 0.0
28	0.1000E 01	0.1000E 03	0.2000E 01	0.1000E 01	0.1000E 01	13	17	3 -90.00	1 0.0	1 0.0
29	0.1000E 01	0.1000E 03	0.2000E 01	0.1000E 01	0.1000E 01	17	21	3 -90.00	1 0.0	1 0.0
30	0.1000E 01	0.1000E 03	0.2000E 01	0.1000E 01	0.1000E 01	21	25	3 -90.00	1 0.0	1 0.0
31	0.1000E 01	0.1000E 03	0.2000E 01	0.1000E 01	0.1000E 01	2	6	3 -90.00	1 0.0	1 0.0
32	0.1000E 01	0.1000E 03	0.2000E 01	0.1000E 01	0.1000E 01	6	10	3 -90.00	1 0.0	1 0.0
33	0.1000E 01	0.1000E 03	0.2000E 01	0.1000E 01	0.1000E 01	10	14	3 -90.00	1 0.0	1 0.0
34	0.1000E 01	0.1000E 03	0.2000E 01	0.1000E 01	0.1000E 01	14	18	3 -90.00	1 0.0	1 0.0
35	0.1000E 01	0.1000E 03	0.2000E 01	0.1000E 01	0.1000E 01	18	22	3 -90.00	1 0.0	1 0.0
36	0.1000E 01	0.1000E 03	0.2000E 01	0.1000E 01	0.1000E 01	22	26	3 -90.00	1 0.0	1 0.0
37	0.1000E 01	0.1000E 03	0.2000E 01	0.1000E 01	0.1000E 01	4	8	3 -90.00	1 0.0	1 0.0
38	0.1000E 01	0.1000E 03	0.2000E 01	0.1000E 01	0.1000E 01	8	12	3 -90.00	1 0.0	1 0.0
39	0.1000E 01	0.1000E 03	0.2000E 01	0.1000E 01	0.1000E 01	12	16	3 -90.00	1 0.0	1 0.0
40	0.1000E 01	0.1000E 03	0.2000E 01	0.1000E 01	0.1000E 01	16	20	3 -90.00	1 0.0	1 0.0
41	0.1000E 01	0.1000E 03	0.2000E 01	0.1000E 01	0.1000E 01	20	24	3 -90.00	1 0.0	1 0.0
42	0.1000E 01	0.1000E 03	0.2000E 01	0.1000E 01	0.1000E 01	24	28	3 -90.00	1 0.0	1 0.0
43	0.1000E 01	0.1000E 03	0.2000E 01	0.1000E 01	0.1000E 01	3	7	3 -90.00	1 0.0	1 0.0
44	0.1000E 01	0.1000E 03	0.2000E 01	0.1000E 01	0.1000E 01	7	11	3 -90.00	1 0.0	1 0.0
45	0.1000E 01	0.1000E 03	0.2000E 01	0.1000E 01	0.1000E 01	11	15	3 -90.00	1 0.0	1 0.0
46	0.1000E 01	0.1000E 03	0.2000E 01	0.1000E 01	0.1000E 01	15	19	3 -90.00	1 0.0	1 0.0
47	0.1000E 01	0.1000E 03	0.2000E 01	0.1000E 01	0.1000E 01	19	23	3 -90.00	1 0.0	1 0.0
48	0.1000E 01	0.1000E 03	0.2000E 01	0.1000E 01	0.1000E 01	23	27	3 -90.00	1 0.0	1 0.0

JOINT DISPLACEMENTS

JOINT	X	Y	Z	TH X	TH Y	TH Z
1	-0.70385843D-07	-0.59663021D-04	0.22890559D-01	-0.16846438D-04	0.43652443D-09	-0.74179944D-10
2	-0.70385843D-07	-0.59663406D-04	0.22890517D-01	-0.16846408D-04	0.43652452D-09	-0.74179872D-10
3	-0.19266119D-07	-0.59663029D-04	0.22888851D-01	-0.16832243D-04	0.43653172D-09	-0.41326023D-10
4	-0.19266119D-07	-0.59663397D-04	0.22888893D-01	-0.16832273D-04	0.43653717D-09	-0.41326095D-10
5	-0.59271917D-07	-0.57785499D-04	0.19289743D-01	-0.27386212D-04	0.34420241D-09	-0.11013125D-09
6	-0.59271917D-07	-0.57785870D-04	0.19289709D-01	-0.27386161D-04	0.34420257D-09	-0.11013119D-09
7	-0.16242125D-07	0.57785507D-04	0.19289708D-01	-0.27378683D-04	0.34420892D-09	-0.57121104D-10
8	-0.16242125D-07	0.57785862D-04	0.19289742D-01	-0.27378734D-04	0.34420876D-09	-0.57121172D-10
9	-0.46372471D-07	-0.52760779D-04	0.15100763D-01	-0.28618504D-04	0.25835929D-09	-0.11571007D-09
10	-0.46372471D-07	-0.52761118D-04	0.15100737D-01	-0.28618451D-04	0.25835945D-09	-0.11571000D-09
11	-0.11795802D-07	-0.52760786D-04	0.15100737D-01	-0.28620009D-04	0.25835610D-09	-0.63952891D-10
12	-0.11795802D-07	-0.52761110D-04	0.15100763D-01	-0.28620061D-04	0.25835595D-09	-0.63952962D-10
13	-0.33173156D-07	-0.44428386D-04	0.10852209D-01	-0.28573774D-04	0.17388420D-09	-0.11613950D-09
14	-0.33173156D-07	-0.44428671D-04	0.10852209D-01	-0.28573723D-04	0.17388435D-09	-0.11613943D-09
15	-0.70064824D-08	0.44428393D-04	0.10852209D-01	-0.28573400D-04	0.17388488D-09	-0.63970155D-10
16	-0.70064824D-08	-0.44428664D-04	0.10852227D-01	-0.28573451D-04	0.17388472D-09	-0.63970223D-10
17	-0.20062255D-07	-0.32773802D-04	0.66360875D-02	-0.27971306D-04	0.90703268D-10	-0.11391819D-09
18	-0.20062255D-07	-0.32774013D-04	0.66360782D-02	-0.27971259D-04	0.90703410D-10	-0.11391813D-09
19	-0.24612158D-08	0.32773808D-04	0.66360782D-02	-0.27971326D-04	0.90703287D-10	-0.59095441D-10
20	-0.24612158D-08	0.32774006D-04	0.66360875D-02	-0.27971373D-04	0.90703144D-10	-0.59095506D-10
21	-0.76209515D-08	-0.17841328D-04	0.26187596D-02	-0.24597394D-04	0.14940637D-10	-0.10011803D-09
22	-0.76209516D-08	-0.17841444D-04	0.26187571D-02	-0.24597365D-04	0.14940727D-10	-0.10011800D-09
23	0.68455811D-09	0.17841333D-04	0.26187571D-02	-0.24597352D-04	0.14940750D-10	-0.38609806D-10
24	0.68455813D-09	0.17841439D-04	0.26187596D-02	-0.24597381D-04	0.14940661D-10	-0.38609845D-10

MEMBER	THRUST	TORQUE	MY(+)	MZ(+)	MY(-)	MZ(-)	JOINTS	
1	-0.66755262D-11	0.71583099D-05	0.30943149D-04	-0.12659667D-03	0.30943095D-04	-0.12659671D-03	1	2
2	-0.31228847D-11	0.71583108D-05	0.30966053D-04	-0.67758446D-04	0.30965999D-04	-0.67758489D-04	4	3
3	-0.12011957D-10	0.12273122D-04	-0.45939921D-06	-0.19154808D-03	-0.45942070D-06	-0.19154812D-03	5	6
4	-0.62607521D-11	0.12273124D-04	-0.44749514D-06	-0.96416993D-04	-0.44806818D-06	-0.96416934D-04	8	7
5	-0.12677556D-10	0.12566456D-04	-0.83365157D-06	-0.20217779D-03	-0.83374647D-06	-0.20217795D-03	9	11
6	-0.70781658D-11	0.12566457D-04	-0.83966950D-06	-0.10928382D-03	-0.83976458D-06	-0.10928386D-03	12	13
7	-0.12841359D-10	0.12298558D-04	-0.32347997D-06	-0.20391067D-03	-0.32357285D-06	-0.20391071D-03	13	14
8	-0.71903464D-11	0.12298557D-04	-0.32253347D-06	-0.11026018D-03	-0.32262653D-06	-0.11026022D-03	16	15
9	-0.12896235D-10	0.11296458D-04	-0.28856431D-06	-0.20125050D-03	-0.28857287D-06	-0.20125054D-03	17	18
10	-0.69100059D-11	0.11296458D-04	-0.28858643D-06	-0.10280660D-03	-0.28859050D-06	-0.10280664D-03	20	19
11	-0.11796095D-10	0.70627300D-05	-0.17500865D-04	-0.17812947D-03	-0.17500919D-04	-0.17812994D-03	21	22
12	-0.49543365D-11	0.70627299D-05	-0.17500823D-04	-0.67602043D-04	-0.17500876D-04	-0.17602066D-04	24	23
13	-0.49989494D 00	0.78957176D-05	-0.13449080D-03	0.28158715D 02	-0.13449845D-03	0.28167214D 02	4	1
14	-0.49989494D 00	0.78957176D-05	-0.13449064D-03	0.28158661D 02	-0.13449828D-03	0.28167160D 02	2	5
15	-0.15820394D-03	0.12728117D-04	-0.15496416D-03	0.47205924D 02	-0.15496797D-03	0.47210410D 02	8	5
16	-0.15820416D-03	0.12728117D-04	-0.15496388D-03	0.47205832D 02	-0.15496769D-03	0.47210318D 02	6	7
17	-0.64205114D-04	0.12420519D-04	-0.15733718D-03	0.49615782D 02	-0.15733510D-03	0.49614753D 02	7	11
18	-0.64205241D-04	0.12420519D-04	-0.15733718D-03	0.49615687D 02	-0.15733489D-03	0.49614753D 02	10	12
19	-0.13355550D-04	0.12520873D-04	-0.15800785D-03	0.49832887D 02	-0.15800788D-03	0.49833081D 02	11	16
20	-0.13355568D-04	0.12520873D-04	-0.15800757D-03	0.49832979D 02	-0.15800816D-03	0.49833081D 02	14	15
21	-0.27809752D-05	0.13157394D-04	-0.15355290D-03	0.49168571D 02	-0.15355282D-03	0.49168531D 02	15	19
22	-0.27809804D-05	0.13157394D-04	-0.15355268D-03	0.49168486D 02	-0.15355257D-03	0.49168446D 02	18	24
23	-0.60199072D-06	0.14761976D-04	-0.12260597D-03	0.43633004D 02	-0.12260598D-03	0.43632990D 02	19	23
24	-0.60199129D-06	0.14761976D-04	-0.12260581D-03	0.43632951D 02	-0.12260582D-03	0.43632990D 02	22	23
25	-0.56325664D 00	-0.14126130D-04	0.28167221D 02	0.45063231D-04	0.21843357D 02	-0.23492446D-04	1	5
26	-0.15074160D 01	-0.19663360D-04	0.25367066D 02	0.30893598D-04	0.24627691D 02	-0.27546309D-04	5	9
27	-0.24997180D 01	-0.20308106D-04	0.24981169D 02	0.29528021D-04	0.25014006D 02	-0.29270363D-04	9	13
28	-0.34963752D 01	-0.20424495D-04	0.24819179D 02	0.28943190D-04	0.25180660D 02	-0.30276693D-04	13	17
29	-0.44797404D 01	-0.20753903D-04	0.23987888D 02	0.28024241D-04	0.26012229D 02	-0.30364225D-04	17	21
30	-0.53523984D 01	-0.22328610D-04	0.17620790D 02	0.18055070D-04	0.32379226D 02	-0.78125891D-04	21	25
31	-0.56326062D 00	-0.14126131D-04	0.28167131D 02	0.45063350D-04	0.21843301D 02	-0.23492571D-04	2	6
32	-0.15074258D 01	-0.19663360D-04	0.25367004D 02	0.30893724D-04	0.24627630D 02	-0.27546437D-04	6	10
33	-0.24997340D 01	-0.20308106D-04	0.24981122D 02	0.29528150D-04	0.25013974D 02	-0.29270490D-04	10	14
34	-0.34963975D 01	-0.20422495D-04	0.24819122D 02	0.28944035D-04	0.25180600D 02	-0.30276815D-04	14	18
35	-0.44797707D 01	-0.20753902D-04	0.23987835D 02	0.28024237D-04	0.26012171D 02	-0.36304317D-04	18	22
36	-0.53524332D 01	-0.22328609D-04	0.17620781D 02	0.18055124D-04	0.32379200D 02	-0.78125921D-04	22	26
37	-0.56326052D 00	-0.14122567D-04	0.28158722D 02	-0.29394420D-04	0.21830845D 02	-0.39016488D-04	4	8
38	-0.15074252D 01	-0.19658800D-04	0.25375091D 02	-0.27170543D-04	0.24630295D 02	-0.31269617D-04	8	12
39	-0.24997339D 01	-0.20308100D-04	0.24985499D 02	0.29394142D-04	0.25013465D 02	-0.29404498D-04	12	16
40	-0.34963974D 01	-0.20422353D-04	0.24819526D 02	-0.31072841D-04	0.25180773D 02	-0.26018578D-04	16	20
41	-0.44797703D 01	-0.20753929D-04	0.23987810D 02	-0.38309975D-04	0.26012205D 02	-0.36507569D-04	20	24
42	-0.53524316D 01	-0.22328605D-04	0.17620806D 02	0.29539314D-04	0.32373234D 02	-0.39016363D-04	24	28
43	-0.56325625D 00	-0.14122567D-04	0.28158665D 02	-0.27170417D-04	0.24630234D 02	-0.31269489D-04	3	7
44	-0.15074161D 01	-0.19658804D-04	0.25375030D 02	-0.29394031D-04	0.24630234D 02	-0.29404371D-04	7	11
45	-0.24997181D 01	-0.20308106D-04	0.24985440D 02	-0.31077215D-04	0.25013463D 02	-0.28147887D-04	11	15
46	-0.34963754D 01	-0.20422353D-04	0.24819469D 02	-0.38308968D-04	0.25180713D 02	-0.26018487D-04	15	19
47	-0.44797424D 01	-0.20753928D-04	0.23987762D 02	-0.38309868D-04	0.26012147D 02	-0.26018487D-04	19	23
48	-0.53524000D 01	-0.22328604D-04	0.17620797D 02	-0.59673423D-04	0.32379208D 02	-0.36507539D-04	23	27

P.5 THE MESH METHOD FOR PLANE FRAMES

The program is a direct application of the material of Chapter 8 to the analysis of an arbitrary plane frame subjected to external joint loads. In addition to the usual quantities required for the description of a frame, this program requires as input a mesh description; it computes joint displacements and member forces.

General Notes

The program follows Section 7.3 but unfortunately this section leaves many things unsaid. It is the intent of these notes to use Program P.5 as a vehicle to clarify the mesh method.

The first section of the program computes a set of forces $\bar{F}$ which satisfy equilibrium with respect to the applied joint loads P. Prior to this, the structure is reduced to a statically determinate structure by removing members. This statically determinate structure consists of a tree associated with each support. The members not included in this tree (the "links") identify a set of redundants. In order to include the effect of the ground, fictitious members are added in the form of a tree which contains all the support points. (In the sample program this tree degenerates into a single member which is shown dashed.) The members of the tree have been arranged in a special fashion (which is always possible) so that it is possible to proceed through the nodes in order using statics and compute:

1. a. The force at the positive end of member 1 (node 1)
   b. The member force in member 1
   c. The force at the negative end of member 1
2. a. The force at the positive end of member 2 (node 2)
   b. The member force in member 2
   c. The force at the negative end of member 2.
      etc.

The second part of the program generates the system matrix and the right hand side. Note that the links have been listed after the members of the tree, that they occur first and positively in the mesh description, and that the fictitious members are listed last. As in the other programs, the contributions of each of the members are added to form the system matrix but here each member is not limited to contributing four terms.

It is possible to generate a force system which satisfies equilibrium by assuming an arbitrary force in one of the members of a mesh and then proceeding around the mesh using the equations of statics to find the forces in the other members. This fact implies an arbitrariness in F_M which is eliminated in this program by assuming that when the only forces present are due to $(F_M)_I$, that the force in link I is $(F_M)_I$.

Flow Chart

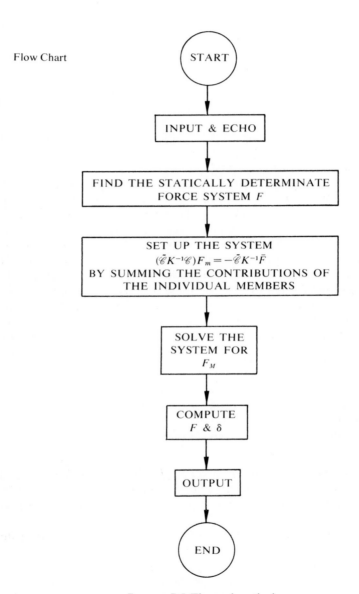

Program P.5. The mesh method.

The third part of the program solves the system matrix for F_M. The fourth part computes the branch force matrix F and the final part computes the joint displacements.

Step by Step Explanation of Program P.5

Step 1   Zero the arrays:
SK(I, J) — member flexibility K_i^{-1}
SN(I, J) — matrix $N_i^\pm$
SNI(I, J) — inverse matrix $(N_i^\pm)^{-1}$
R(I, J) — rotation matrix R_i

Step 2   Define:
E — Young's modulus to 30×10^6
PI — conversion factor, degrees to radians
MXBL — maximum number of members in a single loop set to 6

Step 3   Input:
NB — number of members
NN — total number of nodes (including supports)
NS — number of supports
NL — number of meshes
NLD — number of loaded nodes

Step 4   Zero the array:
NJT(I) — an index to indicate a loaded node
BF(K, J) — the member (branch) force matrix F_K

Step 5   Input:
A(K) — area of member K
AL(K) — length of member K
SI(K) — moment of inertia of member K
TH(K) — the angle of orientation, Φ_K, of member K
LP(K) — the positive end of member K
MI(K) — the negative end of member K
BF(K, J) — applied joint loads are initially stored in this array.

Step 6   Compute and output the statically determinate force system $\bar{F}$ by working backward from the tips of the trees as described in the explanation of the mesh method. At the end of this step $\bar{F}$ appears in the array $BF(I, J)$.

Step 7   Zero the arrays:
   $P(I)$ — the right hand side
   $C(I, J)$ — the system matrix

Step 8   Define $NB1$, the number of real branches.

Step 9   Compute the contribution of each member to the system matrix C and the right hand side P.

Step 10   Solve the system matrix.

Step 11   Compute the member force matrix F from the mesh forces F_M. (At the end of step 10, F_M appears as the array $P(I)$.)

Step 12   Compute joint displacements. This is done in the inverse order of step 6, starting from the supports.

Fortran Program

```
C     THE MESH METHOD FOR PLANE FRAMES
      DOUBLE PRECISION C(100,100),P(100),SK(3,3),SN(3,3),SNI(3,3),
     1 R(3,3), BF(25,3),D(3),TH(25),CB(3,3,6),E1(3,3),E2(3,3),THA
      DIMENSION NJT(25),A(25), AL(25),SI(25),LP(25),MI(25),MESH(25,6),
     1 NLK(6)
      DO 6I=1,3
      DO 6J=1,3
      SK(I,J)=0.
      SN(I,J)=0.
      SNI(I,J)=0.
    6 R(I,J)=.0
      E=30000000.
      PI=3.14159     /180.
      MXBL=6
      R(3,3)=1.
  888 READ          (5,2)NB,NN,NS ,NL,NLD
    2 FORMAT (6(I3,3X))
      NNS=NN-NS
      N=3*NL
      DO1204 I=1,NNS
 1204 NJT(I)=0
      WRITE         (6,901)
  901 FORMAT(18H1MEMBER PROPERTIES /11X,4HAREA,15X,6HLENGTH,16X,1HI,17X,
     1 5HANGLE,14X,5H+ END,5X,5H- END)
      DO26 K=1,NB
      DO 5 J=1,3
    5 BF(K,J)=0.
      READ          (5,13)A(K),AL(K),SI(K),TH(K),LP(K),MI(K)
   26 WRITE         (6,900)K, A(K),AL(K),SI(K),TH(K),LP(K),MI(K)
  900 FORMAT(I4,4E20.8,2I10)
      DO 12 I=1,NLD
      READ          (5,2)K
      NJT(K)=K
   12 READ          (5,3)(BF(K,J),J=1,3)
    3 FORMAT(4(E10.2,2X))
      WRITE         (6,4)(NJT(I),(BF(I,J),J=1,3),I=1,NNS)
    4 FORMAT(20H1APPLIED JOINT LOADS///5X,5HJOINT,18X,2HPX,18X,2HPY,19X,
     21HM//(I10,3E20.8))
   13 FORMAT(4(E10.2,2X),2(3X,I3))
      KK=1
      ZK=-1.
      ZKI=1.
      NODE=0
  151 NODE=NODE+1
      IF(NJT(NODE))7,10,7
    7 K=NODE
      GO TO 113
   34 DO 11 I=1,3
```

```
      D(I)=0.
      DO 11 J=1,3
   11 D(I)=D(I)+R(I,J)*BF(NODE,J)
      DO 14 I=1,3
      BF(NODE,I)=0.
      DO 14 J=1,3
   14 BF(NODE,I)=BF(NODE,I)+SNI(J,I)*D(J)
      IF(MI(NODE)-NNS)15,15,10
   15 DO 16 I=1,3
      D(I)=0.
      DO 16 J=1,3
   16 D(I)=D(I)+SN(J,I)*BF(NODE,J)
      K=MI(NODE)
      DO 17 I=1,3
      DO 17 J=1,3
   17 BF(K      ,I)=BF(K      ,I)-R(J,I)*D(J)
      NJT(K)=K
   10 IF(NODE-NNS)151,9,9
    9 WRITE             (6,1)(I,(BF(I,J),J=1,3),I=1,NB)
    1 FORMAT(22H1INITIAL BRANCH FORCES///4X,6HBRANCH,14X,6HTHRUST,18X,2HM+
     2M+,18X,2HM-//(I10,3E20.8))
      READ              (5,2)((MESH(I,J),J=1,6),I=1,NL)
      WRITE             (6,902)(I,(MESH(I,J),J=1,6),      I=1,NL)
  902 FORMAT(17H1MESH DESCRIPTION///(5H MESH,I3,11H...BRANCHES,6I5))
      DO 904 I=1,N
      P(I)=0.
      DO 904 J=1,N
  904 C(I,J)=0.
      NB1=NB+1- NS
      KK=2
      I1=0
  152 I1=I1+1
      N1=      1
      LL=1
      J1=0
  154 J1=J1+1
      DO 42 K1=1,MXBL
      IF(IABS (MESH(J1,K1))-I1)42,143,42
  143 K11=K1
      GO TO 43
   42 CONTINUE
      GO TO 41
   43 DO 44 I=1,3
      DO 45 J=1,3
   45 CB(I,J,N1)=0.
   44 CB(I,I,N1)=1.
      NLK(N1)=J1
      K1=0
```

COMPUTE F

```
156 K1=K1+1
319 FORMAT(6I10)
    M1=MESH(J1,K1)
    K=IABS (M1)
    ZK=ISIGN (1,M1)
    ZKI=-ZK
    GO TO 113
157 IF(K1-1) 72,48,72
 72 DO 71 I=1,3
    DO 71 J=1,3
    E1(I,J)=0.
    DO 71 K=1,3
 71 E1(I,J)=E1(I,J)+R(I,K)*E2(K,J)
 51 DO 50 I=1,3
    DO 50 J=1,3
    CB(I,J,N1)=0.
    DO 50 K=1,3
 50 CB(I,J,N1)=CB(I,J,N1)-SNI(K,I)*E1(K,J)
 48 DO 49 I=1,3
    DO 49 J=1,3
    E1(I,J)=0.
    DO 49 K=1,3
 49 E1(I,J)=E1(I,J)+SN(K,I)*CB(K,J,  N1)
    DO 73 I=1,3
    DO 73 J=1,3
    E2(I,J)=0.
    DO 73 K=1,3
 73 E2(I,J)=E2(I,J)+R(K,I)*E1(K,J)
 74 IF(K1-K11)156,53,53
 53 N1=N1+1
 41 IF(J1-NL)154,155,155
155 N1=N1-1
    K=I1
    LL=2
    GO TO 37
 57 DO 55 J1=1,N1
    J11=3*NLK(J1)-3
    DO 56 I=1,3
    DO 56 J=1,3
    E1(I,J)=0.
    DO 56 K=1,3
 56 E1(I,J)=E1(I,J)+CB(K,I,J1)*SK(K,J)
    DO 58 K1=1,N1
    K11=3*NLK(K1)-3
    DO 59 I=1,3
    M1=J11+I
    DO 59 J=1,3
    N2=K11+J
```

GENERATE SYSTEM MATRIX & RIGHT HAND SIDE

```
      DO 59 K=1,3
   59 C(M1,N2)=C(M1,N2)+E1(I,K)*CB(K,J,K1)
   58 CONTINUE
      IF(NJT(I1))60,55,60
   60 DO161 I=1,3
      M1=J11+I
      DO161 J=1,3
  161 P(M1)=P(M1)-E1(I,J)*BF(I1,J)
   55 CONTINUE
   40 IF(I1-NB1)152,153,153
  153 M=N-1
      DO 91 I=1,M
      L=I+1
      DO 91 J=L,N
      IF (C(J,I)) 93,91,93
   93 DO 92 K=L,N
   92 C(J,K)=C(J,K)-C(I,K)*C(J,I)/C(I,I)
      P(J)=P(J)-P(I)           *C(J,I)/C(I,I)
   91 CONTINUE
      P (N)=P(N)/C(N,N)
      DO 94 I=1,M
      K=N-I
      L=K+1
      DO 95 J=L,N
   95 P(K)=P(K)-P (J)*C(K,J)
   94 P (K)=P(K)/C(K,K)
      I1=0
  251 I1=I1+1
      DO 63 I=1,3
      J=3*I1-3+I
   63 D(I)=P(J)
      J1=0
  253 J1=J1+1
      IF(MESH(I1,J1)) 79,61,79
   79 M1=MESH(I1,J1  )
      K=IABS (M1)
      ZKI=-ISIGN (1,M1)
      ZK=-ZKI
      KK=3
      GO TO 113
   68 IF( J1-1) 77,78,77
   77 DO 75 I=1,3
      E1(1,I)=0.
      DO 75 J=1,3
   75 E1(1,I)=E1(1,I)-R(I,J)*D(J)
      DO 69 I=1,3
      D(I)=0.
      DO 69 J=1,3
```

SOLVE SYSTEM FOR F_M

COMPUTE F

```
  69 D(I)=D(I)+SNI(J,I)*E1(1,J)
  78 DO 64 I=1,3
  64 BF(K,I)=D(I)+BF(K,I)
  67 DO 66 I=1,3
     E1(1,I)=0.
     DO 66 J=1,3
  66 E1(1,I)=E1(1,I)+SN(J,I)*D(J)
     DO 76 I=1,3
     D(I)=0.
     DO 76 J=1,3
  76 D(I)=D(I)+R(J,I)*E1(1,J)
  62 IF(J1-MXBL)253,61,61
  61 IF(I1-NL)251,252,252
 252 WRITE            (6,70)(I,(BF(I,J),J=1,3),I=1,NB)
  70 FORMAT(22H1  FINAL  BRANCH FORCES///4X,6HBRANCH,14X,6HTHRUST,18X,2HM+
    2M+,18X,2HM-//(I10,3E20.8))
     WRITE(6,203)
 203 FORMAT(20H1JOINT DISPLACEMENTS ///6H JOINT,18X,2HDX,18X,2HDY,18X,
    2 2HTH//)
     KK=2
     ZK=-1.
     ZKI=1.
     LL=3
     I1=0
 254 I1=I1+1
     K=NNS-I1+1
     GO TO 113
 201 IF(MI(K)-NNS) 204,204,205
 204 YK=1.
     GO TO 206
 205 YK=0.
 206 J1=MI(K)
     DO 207 I=1,3
     E1(1,I)=0.
     DO 207 J=1,3
 207 E1(1,I)=E1(1,I)+R(I,J)*BF(J1,J)*YK
     DO 202 I=1,3
     D(I)=0.
     DO 202 J=1,3
 202 D(I)=D(I)+SK(I,J)*BF(K,J)-SN(I,J)*E1(1,J)
     DO 208 I=1,3
     E1(3,I)=0.
     DO 208 J=1,3
 208 E1(3,I)=E1(3,I)+SNI(I,J)*D(J)
     DO 209 I=1,3
     BF(K,I)=0.
     DO 209 J=1,3
 209 BF(K,I)=BF(K,I)+R(J,I)*E1(3,J)
```

COMPUTER 8

```
      WRITE(6,210)K,(BF(K,I),I=1,3)
210 FORMAT(I6,3E20.8)
200 IF(I1-NNS)254,888,888
113 THA=TH(K)*PI
      R(1,1)=DCOS(THA)
      R(2,2)=R(1,1)
      R(1,2)=DSIN(THA)
      R(2,1)=-R(1,2)
352 SN(1,1)=ZK
      SN(2,2)=-ZK/AL(K)
      SN(3,2)=-ZK/AL(K)
      IF(ZK)35,35,36
 36 SN(2,3)=1.
      SN(3,3)=0.
      GO TO 351
 35 SN(2,3)=0.
      SN(3,3)=1.
351 CONTINUE
 30 SNI(1,1)=ZKI
      SNI(3,2)=ZKI
      SNI(3,3)=-ZKI
      IF(ZKI)31,32,32
 31 SNI(2,2)=AL(K)
      SNI(2,3)=0.
      GO TO 33
 32 SNI(2,2)=0.
      SNI(2,3)=-AL(K)
 33 GO TO (34,37,68          ),KK
 37 SK(1,1)=AL(K)/(E*A(K))
      SK(2,2)=AL(K)/(E*3.*SI(K))
      SK(3,3)=SK(2,2)
      SK(2,3)=-.5*SK(2,2)
      SK(3,2)=SK(2,3)
      GO TO (157,57,201          ),LL
      END
```

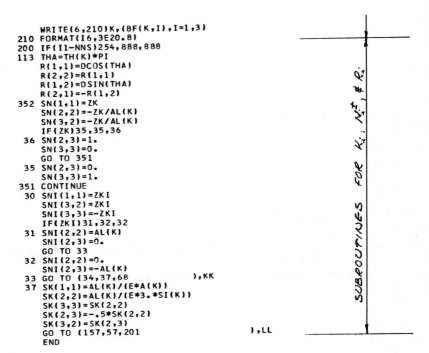

Sample Problem

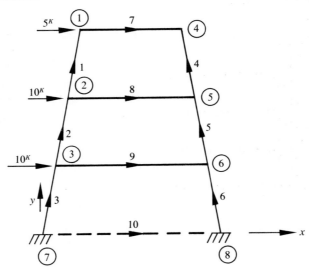

Program P.5. Sample problem.

Card Input

```
/*
//GO.SYSIN DD*
10      8       2       3       3
    3.00E3      181.40E0    1.00E3      .82.87E0    1       2
    3.00E3      181.40E0    2.00E3      .82.87E0    2       3
    3.00E3      217.68E0    3.00E3      .82.87E0    3       7
    3.00E3      181.40E0    1.00E3      97.13E0     4       5
    3.00E3      181.40E0    2.00E3      97.13E0     5       6
    3.00E3      217.68E0    3.00E3      97.13E0     6       8
    3.00E3      144.00E0    1.00E3      .00.00E0    4       1
    3.00E3      189.00E0    1.00E3      .00.00E0    5       2
    3.00E3      234.00E0    1.00E3      .00.00E0    6       3
                288.                                8       7
  1
    5.00E3      000.00E0    0.00E3
  2
    10.00E3     000.00E0    0.00E3
  3
    10.00E3     000.00E0    0.00E3
    7   -4      -8      1
    8   -5      -9      2
    9   -6      -10     3
/*
```

Computer Output

MEMBER PROPERTIES

	AREA	LENGTH	I	ANGLE	+ END	- END
1	0.30000000E 04	0.18139999E 03	0.10000000E 04	0.82870000D 02	1	2
2	0.30000000E 04	0.18139999E 03	0.20000000E 04	0.82870000D 02	2	3
3	0.30000000E 04	0.21767999E 03	0.30000000E 04	0.82870000D 02	3	7
4	0.30000000E 04	0.18139999E 03	0.10000000E 04	0.97130000D 02	4	5
5	0.30000000E 04	0.21767999E 03	0.20000000E 04	0.97130000D 02	5	6
6	0.30000000E 04	0.14400000E 03	0.30000000E 04	0.0	5	1
7	0.30000000E 04	0.18900000E 03	0.10000000E 04	0.0	6	2
8	0.30000000E 04	0.23400000E 03	0.10000000E 04	0.0	8	3
9	0.30000000E 04	0.28800000E 03	0.10000000E 04	0.0		7
10	0.0		0.0			

APPLIED JOINT LOADS

JOINT	PX	PY	M
1	0.50000000D 04	0.0	0.0
2	0.10000000D 05	0.0	0.0
3	0.10000000D 05	0.0	0.0
0	0.0	0.0	0.0
0	0.0	0.0	0.0

INITIAL BRANCH FORCES

BRANCH	THRUST	M+	M-
1	0.62061370D 03	0.0	0.89998602D 06
2	0.18618411D 04	-0.89998602D 06	0.35999437D 07
3	0.31030685D 04	-0.35999437D 07	0.89998581D 07
4	0.0	0.0	0.0
5	0.0	0.0	0.0
6	0.0	0.0	0.0
7	0.0	0.0	0.0
8	0.0	0.0	0.0
9	0.0	0.0	0.0
10	0.0	0.0	0.0

MESH DESCRIPTION

MESH	1...BRANCHES	7	-4	-8	1	0	0
ESH	2...BRANCHES	8	-5	-9	2	0	0
MESH	3...BRANCHES	9	-6	-10	3	0	0

FINAL BRANCH FORCES

BRANCH	THRUST	M+	M-
1	0.45504487D 04	0.30755715D 06	0.46121429D 05
2	0.13585137D 05	0.75483526D 06	0.30765345D 06
3	0.22289855D 05	0.64479898D 06	0.149028730 07
4	-0.45505784D 04	0.30765946D 06	0.46219095D 05
5	-0.13585589D 05	0.75511484D 06	0.30807075D 06
6	-0.22290016D 05	0.64509043D 06	0.14903585D 07
7	-0.25005474D 04	-0.30765946D 06	-0.30768952D 06
8	-0.50013639D 04	-0.80133393D 06	-0.80135238D 06
9	-0.49988877D 04	-0.95316118D 06	-0.95322946D 06
10	0.12500792D 05	-0.149035850 07	0.75095708D 07

JOINT DISPLACEMENTS

JOINT	DX	DY	TH
6	0.20337131D 00	0.25384633D-01	-0.10222109D-02
5	0.37476512D 00	0.46796056D-01	-0.34642932D-03
4	0.39808190D 00	0.49703423D-01	0.44399201D-03
3	C.20338418D 00	-0.2538701 1D-01	-0.10224772D-02
2	C.37477551D 00	-0.46798770D-01	-0.34648743D-03
1	0.39808586D 00	-0.497054 21D-01	0.443919 86D-03

P.6 A PROGRAM FOR LARGE SPACE TRUSSES

General Comments

This program is described in some detail in Appendix A.5. It is on the whole beyond the scope of this book but is included here to indicate to the reader that it is a relatively simple matter to improve the computational efficiency of a program such as a space truss program to the point where the solution of structures which require thousands of equations becomes routine.

Step by Step Explanation of Program P.6

Step 1 Rewind tape 12. the scratch tape. Set E. Young's modulus to 30×10^6 psi.

Step 2 Input problem parameters,
  NNS — number of movable joints
   NB — number of bars
  A1(I) — area of bar I
  AL(I) — length of bar I
AN(I, 1) — x component of the unit vector $\mathbf{n}_I$
AN(I, 2) — y component of the unit vector $\mathbf{n}_I$
AN(I, 3) — z component of the unit vector $\mathbf{n}_I$
  NP(I) — positive end of bar I
  MI(I) — negative end of bar I
  P(I, 1) — x component of the joint load vector $\mathbf{P}_I$
  P(I, 2) — y component of the joint load vector $\mathbf{P}_I$
  P(I, 3) — z component of the joint load vector $\mathbf{P}_I$

Step 3 Initialize,
  NG — length of the major array $B(NG, 3, 3)$ which is used to store the coefficients of the equations after elimination has been performed.
NSTOR — is the number of unused elements in the array B

Step 4 Generate equation IZ in the array C,
1. Zero the diagonal term of the equation $C(NF, I, J)$
2. Compute the contribution of each bar to equation IZ

Step 5 Eliminate the terms to the left of the diagonal in equation IZ.

Step 6 Multiply the remaining equation by the inverse of the diagonal term.

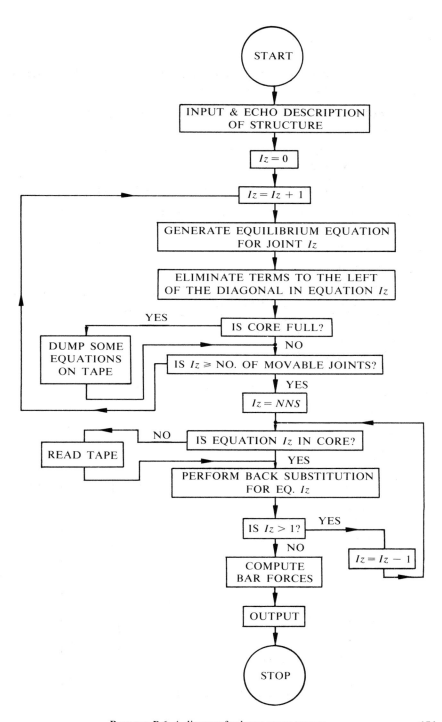

Program P.6. A diagram for large space trusses.   **151**

Step 7   See if there is room for equation *IZ* in the array *B*. If there is not, write some of the terms from *B* onto tape 12.

Step 8   Backsubstitute, reading equations from tape 12 if they are not in array *B*.

Step 9   Write out the joint displacements which appear in the array *P* after the system has been solved.

Step 10   Compute the bar forces from the joint displacements.

Step 11   Check statics by seeing how well the displacements satisfy the equilibrium equations.

Fortran Program

```
C      LARGE SPACE TRUSSES
       REWIND12
       E=30000000.
       DOUBLE PRECISION C(84,3,3),B(1700,3,3),AI(3,3),P(323,3),A11,A12,A13
      13, A21,A22,A23,A31,A32,A33,C11,C12,C13,C21,C22,C23,C31,C32,C33,DET
       DIMENSION A1(1120),AL(1120),AN(1120,3),NP(1120),MI(1120)
      1,N1(323),N3(1700),N2(323),NC(84)
       READ            (5,50)NNS,NB
       DO 1 I=1,NB,2
    1  READ            (5,51)A1(I),AL(I),AN(I,1),AN(I,2),AN(I,3),NP(I),
      1MI(I),A1(I+1),AL(I+1),AN(I+1,1),AN(I+1,2),AN(I+1,3),NP(I+1),
      2MI(I+1)
       DO 2 I=1,NNS,2
    2  READ            (5,56)P(I,1),P(I,2),P(I,3),P(I+1,1),P(I+1,2),
      1 P(I+1,3)
       N2(1)=1
       NG=1700
       NSTOR=NG
       NROW=0
       NF=42
       NF2=2*NF
       NSZ=3
       IZ=0
  900  IZ=IZ+1
       N=0
       DO 221 I=1,NF2
  221  NC(I)=0
       NC(NF)=1
       DO 9 I=1,NSZ
       DO 9 J=1,NSZ
    9  C(NF,I,J)=0.
       M=N2(IZ)
  102  IF(M -NG  )64,64,112
  112  M =M -NG
       GO TO 102
   64  JZ=0
  902  JZ=JZ+1
       IF(NP(JZ)-IZ)5,6,5
    5  IF(MI(JZ)-IZ)4,7,4
    6  I1=MI(JZ)-IZ+NF
       GO TO 8
    7  I1=NP(JZ)-IZ+NF
    8  IF(I1-NNS-NF-1+IZ)13,71,71
   71  I1=NF2
   13  DO 10 I=1,3
       DO 10 J=1,3
       C(I1,I,J)=-AN(JZ,I)*AN(JZ,J)*E*A1(JZ)/AL(JZ)
   10  C(NF,I,J)=C(NF,I,J)-C(I1,I,J)
```

INPUT & ECHO

GENERATE COEFF.

ELIMINATION

152

```
      NC(I1)=1
    4 IF(JZ-NB)902,903,903
  903 NF1=NF-1
      DO 14 JZ=1,NF1
      IF(NC(JZ))65,14,65
   65 NR=JZ+IZ-NF
   15 NZ3=N1(NR)
      I2=JZ
      DO 16 KZ=2,NZ3
      MR =N2(NR)+KZ-1
  104 IF(MR-NG )66,66,114
  114 MR=MR-NG
      GO TO 104
   66 NR1=N3(MR )
   18 I3=MR
      I1=NR1-IZ+NF
      IF(NC(I1))22,21,22
   21 NK=0
      NC(I1)=1
      GO TO 23
   22 NK=1
   23 DO 12 I=1,NSZ
      DO 12 J=1,NSZ
      IF(NK)24,25,24
   25 C(I1,I,J)=0.
   24 DO 12 K=1,NSZ
   12 C(I1,I,J)=C(I1,I,J)-C(I2,I,K)*B(I3,K,J)
   16 CONTINUE
      DO 17 I=1,NSZ
      DO 17 J=1,NSZ
   17 P(IZ,I)=P(IZ,I)-C(I2,I,J)*P(NR,J)
   14 CONTINUE
      A11=C(NF,1,1)
      A12=C(NF,1,2)
      A13=C(NF,1,3)
      A22=C(NF,2,2)
      A23=C(NF,2,3)
      A33=C(NF,3,3)
      A21=C(NF,2,1)
      A31=C(NF,3,1)
      A32=C(NF,3,2)
      C11=A22*A33-A23*A32
      C12=-A21*A33+A31*A23
      C13=A21*A32-A31*A22
      DET=A11*C11+A12*C12+A13*C13
      C21=-A12*A33+A32*A13
      C31=A12*A23-A22*A13
      C32=-A11*A23+A21*A13
```

ELIMINATE

ELIMINATION

```
      C22=A11*A33-A31*A13
      C23=-A11*A32+A31*A12
      C33=A11*A22-A21*A12
      AI(1,1)=C11/DET
      AI(1,2)=C21/DET
      AI(1,3)=C31/DET
      AI(2,3)=C32/DET
      AI(2,1)=C12/DET
      AI(3,1)=C13/DET
      AI(3,2)=C23/DET
      AI(2,2)=C22/DET
      AI(3,3)=C33/DET
      I3=M
      NF3=NF2-1
      DO 28 JZ=NF,NF3
      IF(NC(JZ))26,28,26
   26 N3(I3)=JZ+IZ-NF
      DO 30 I=1,NSZ
      DO 30 J=1,NSZ
      B(I3,I,J)=0.
      DO 30 K=1,NSZ
   30 B(I3,I,J)=B(I3,I,J)+AI(I,K)*C(JZ,K,J)
      NSTOR=NSTOR-1
   73 IF(NSTOR) 72,72,74
   72 NROW=NROW+1
      NSTOR=NSTOR+N1(NROW)
      MB=N2(NROW)
   38 I5=MB+N1(NROW)-1
      IF(I5-NG )41,41,42
   41 WRITE    (12)(((B(I4,I,J),I=1,3),J=1,3),N3(I4),I4=MB,I5)
      GO TO 73
   42 I5=I5-NG
      WRITE    (12)(((B(I4,I,J),I=1,3),J=1,3),N3(I4),I4=MB,NG   ),(((
     1 B(I4,I,J),I=1,3),J=1,3),N3(I4),I4=1,I5)
      GO TO 73
   74 I3=I3+1
      N=N+1
  231 IF(I3-NG )28,28,230
  230 I3=I3-NG
      GO TO 231
   28 CONTINUE
      N1(IZ)=N
      N2(IZ+1)=M+N
      IF(M+N-NG)81,81,82
   82 N2(IZ+1)=M+N-NG
   81 DO 33 I=1,NSZ
      C(I,1,1)=0.
      DO 33 J=1,NSZ
```

TAPE STORAGE

ELIMINATION

```
  33 C(I,1,1)=C(I,1,1)+AI(I,J)*P(IZ,J)
     DO 34 I=1,NSZ
  34 P(IZ,I)=C(I,1,1)
   3 IF(IZ-NNS)900,901,901
 901 WRITE(6,50)NROW
     WRITE            (6,54)
     WRITE            (6,53)
     DO 35IY=1,NNS
     IZ=NNS-IY+1
  36 N=N1(IZ)
     IF(IZ-NROW) 75,37,76
  76 I1=N2(IZ)
     GO TO 80
  75 BACKSPACE12
  37 BACKSPACE12
     READ     (12)(((B(I3,I,J),I=1,3),J=1,3),N3(I3),I3=1,N)
     I1=1
  80 IF(IY-1)35,35,46
  46 DO 43   K=2,N
     JZ=K+I1-1
  79 IF(JZ-NG) 77,77,78
  78 JZ=JZ-NG
     GO TO 79
  77 LZ=N3(JZ)
     DO 43 I=1,NSZ
     DO 43 J=1,NSZ
  43 P(IZ,I)=P(IZ,I)-B(JZ,I,J)*P(LZ,J)
  35 WRITE            (6,52)IZ,P(IZ,1),P(IZ,2),P(IZ,3)
     WRITE            (6,55)
     P(NNS+1,1)=0.
     P(NNS+1,2)=0.
     P(NNS+1,3)=0.
     DO 44 I=1,NB
     L1=NP(I)
     L2=MI(I)
     D=0.
     DO 47 J=1,3
  47 D=D+(P(L1,J)-P(L2,J))*AN(I,J)
     D=E*A1(I)*D/AL(I)
     D1=D/A1(I)
  44 WRITE            (6,57) I,D,D1,NP(I),MI(I)
     IZ=NNS+1
     WRITE(6,910)
 910 FORMAT(///13HSTATIC CHECK  )
     DO 301 I=1,3
 301 AI(I,1)=0.
     DO 307 JZ=1,NB
     IF(NP(JZ)-IZ)302,303,302
```

BACK SUBSTITUTION

COMPUTE F1A

```
302  IF(MI(JZ)-IZ)307,304,307
304  Z=-1.
     GO TO 305
303  Z=1.
305  D=0.
     L1=NP(JZ)
     L2=MI(JZ)
     DO 306 J=1,3
306  D=D+(P(L1,J)-P(L2,J))*AN(JZ,J)
     DO 308 I=1,3
308  AI(I,1)=AI(I,1)+E*A1(JZ)*D*Z*AN(JZ,I)/AL(JZ)
307  CONTINUE
300  WRITE            (6,52)IZ,AI(1,1),AI(2,1),AI(3,1)
     STOP
 50  FORMAT(3(I4,3X))
 51  FORMAT(2(F4.2,F4.0,F6.4,F6.4,F6.4,I4,I4))
 52  FORMAT(I4,3X,3(E18.8,3X))
 53  FORMAT(5HJOINT,16X,1HX,21X,1HY,21X,1HZ)
 54  FORMAT(///19HJOINT DISPLACEMENTS)
 55  FORMAT(///40HBAR             FORCE             STRESS,12X,6HJOINTS
  1 )
 56  FORMAT(6(E10.1))
 57  FORMAT(I6,E18.8,E18.8,I8,I8)
     END
```

STATIC CHECK

P.7 ADDING FLEXIBILITIES

This program is a direct application of the material of Section 7.3 to the problem of the approximation of an arbitrary curved beam by straight uniform segments. It requires as input descriptions of all the pieces; it computes first the cantilever flexibility and then the member stiffness.

General Notes

In this program Eq. (7.10) is used to obtain the stiffness of an arbitrarily curved *plane beam* by approximating it by straight uniform pieces. In the example which follows, the results obtained are compared to "exact" results for a semi-circular uniform beam. The formula to be evaluated by this program is

$$c_{i+1} = R_i(N_i^+)^{-1}\{L_i^{-1} + N_i^- R_i C_i \tilde{R}_i \tilde{N}_i^-\}(\tilde{N}_i^+)^{-1}R_i \qquad (7.10)$$

after which K_i is computed for the *entire member* from

$$K_i = (N_i^+ C_i \tilde{N}_i^+)^{-1} \qquad (7.5)$$

Note that $(N_i^+)^{-1}$ and K_i^{-1} can be determined directly to be

$$(N_i^+)^{-1} = \begin{bmatrix} \dfrac{L_i}{A_i E} & 0 & 0 \\[2ex] 0 & \dfrac{L_i}{3EI_i} & \dfrac{-L_i}{6EI_i} \\[2ex] 0 & \dfrac{-L_i}{6EI_i} & \dfrac{L_i}{3EI_i} \end{bmatrix} \cdot \qquad K_i^{-1} = \begin{bmatrix} 1 & 0 & 0 \\ 0 & 0 & -L \\ 0 & 1 & -1 \end{bmatrix};$$

Flow Chart

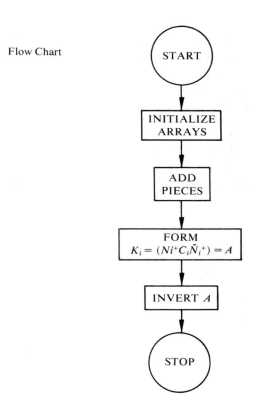

Program P.7. Adding flexibilities.

Step by Step Explanation of Program P.7

Step 1   Define $E = 30 \times 10^6$ psi.

Step 2   Zero the arrays:
   D1(I, J) — scratch array
   D2(I, J) — scratch array
   D3(I, J) — scratch array
   D4(I, J) — scratch array
ANM(I, J) — N_i^-
 ANP(I, J) — N_i^+
   C(I, J) — end of member flexibility
   R(I, J) — rotation matrix R_i
 API(I, J) — $(N_i^+)^{-1}$
  AK(I, J) — K_i^{-1}

Step 3   Input
N — number of segments used to approximate the curved member
AL1 — L_i for the "entire" member

Step 4   For each segment:
a. Read in the segment properties,
    A1 $= A_i$ (area)
    TH $= \theta_i$ (orientation)
    AI1 $= I_i$ (moment of inertia)
    AL $= L_i$ (length)
b. Set up the required arrays
c. Form,
    D1 $= \tilde{R}_i (N_i{}^+)^{-1}$
    D2 $= N_i R_i$
d. Form,
    D3 $= D2 \cdot C_i$
    D4 $= D3 \cdot D2 + K_i{}^{-1}$
e. Form,
    D2 $= D1 \cdot D4$
    C $= D2 \cdot D1$

Step 5   Form,
D1 $= N_i{}^+ C_i$
  A $= D1 \cdot N_i{}^+$

Step 6   Form A^{-1}

Fortran Program

```
      DIMENSION R(3,3),C(3,3), ANP(3,3),ANM(3,3),API(3,3)
      DIMENSION AK(3,3),D1(3,3),D2(3,3),D3(3,3),D4(3,3)
      DIMENSION AI(4,4),A(4,4)
      E=30.E6
      DO 1 I=1,3
      DO 1 J=1,3
      D4(I,J)=0.
      D3(I,J)=0.
      ANM(I,J)=0.
      ANP(I,J)=0.
      C(I,J)=0.
      R(I,J)=0.
      API(I,J)=0.
      AK(I,J)=0.
      D1(I,J)=0.
    1 D2(I,J)=0.
      READ(5,100) N,AL1
  100 FORMAT(I5,E20.8)
  101 FORMAT(4E20.8)
      DO 9 L=1,N
      READ(5,101) A1,TH,AI1,AL
      ANP(1,1)=1.
      ANP(2,3)=1.
      ANP(2,2)=-1./AL
      ANP(3,2)=-1./AL
      ANM(1,1)=-1.
      ANM(3,3)=1.
      ANM(2,2)=1./AL
      ANM(3,2)=1./AL
      API(1,1)=1.
      API(3,2)=1.
      API(3,3)=-1.
      API(2,3)=-AL
      AK(1,1)=AL/(A1*E)
      AK(2,2)=AL/(3.*E*AI1)
      AK(3,3)=AK(2,2)
      AK(2,3)=-AK(2,2)*.5
      AK(3,2)=AK(2,3)
      R(3,3)=1.
      R(1,1)=COS(TH)
      R(1,2)=SIN(TH)
      R(2,2)=R(1,1)
      R(2,1)=-R(1,2)
      DO 3 I=1,3
      DO 3 J=1,3
      DO 3 K=1,3
      D1(I,J)=D1(I,J)+R(K,I)*API(K,J)
    3 D2(I,J)=D2(I,J)+ANM(I,K)*R(K,J)
```

INITIALIZE

ADD PIECES

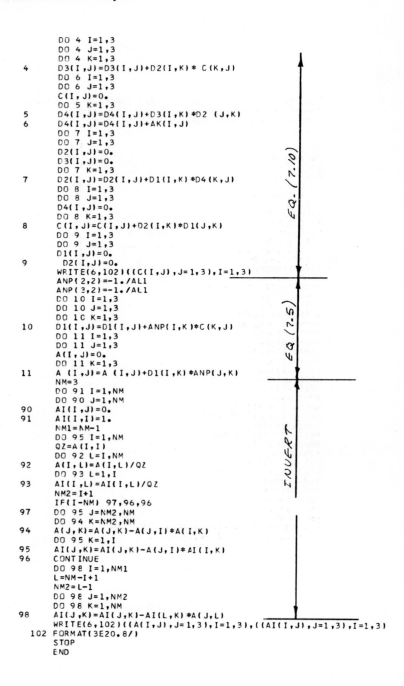

```
        DO 4 I=1,3
        DO 4 J=1,3
        DO 4 K=1,3
4       D3(I,J)=D3(I,J)+D2(I,K)* C(K,J)
        DO 6 I=1,3
        DO 6 J=1,3
        C(I,J)=0.
        DO 5 K=1,3
5       D4(I,J)=D4(I,J)+D3(I,K)*D2 (J,K)
6       D4(I,J)=D4(I,J)+AK(I,J)
        DO 7 I=1,3
        DO 7 J=1,3
        D2(I,J)=0.
        D3(I,J)=0.
        DO 7 K=1,3
7       D2(I,J)=D2(I,J)+D1(I,K)*D4(K,J)
        DO 8 I=1,3
        DO 8 J=1,3
        D4(I,J)=0.
        DO 8 K=1,3
8       C(I,J)=C(I,J)+D2(I,K)*D1(J,K)
        DO 9 I=1,3
        DO 9 J=1,3
        D1(I,J)=0.
9        D2(I,J)=0.
        WRITE(6,102)((C(I,J),J=1,3),I=1,3)
        ANP(2,2)=-1./AL1
        ANP(3,2)=-1./AL1
        DO 10 I=1,3
        DO 10 J=1,3
        DO 1C K=1,3
10      D1(I,J)=D1(I,J)+ANP(I,K)*C(K,J)
        DO 11 I=1,3
        DO 11 J=1,3
        A(I,J)=0.
        DO 11 K=1,3
11      A (I,J)=A (I,J)+D1(I,K)*ANP(J,K)
        NM=3
        DO 91 I=1,NM
        DO 90 J=1,NM
90      AI(I,J)=0.
91      AI(I,I)=1.
        NM1=NM-1
        DO 95 I=1,NM
        QZ=A(I,I)
        DO 92 L=I,NM
92      A(I,L)=A(I,L)/QZ
        DO 93 L=1,I
93      AI(I,L)=AI(I,L)/QZ
        NM2=I+1
        IF(I-NM) 97,96,96
97      DO 95 J=NM2,NM
        DO 94 K=NM2,NM
94      A(J,K)=A(J,K)-A(J,I)*A(I,K)
        DO 95 K=1,I
95      AI(J,K)=AI(J,K)-A(J,I)*AI(I,K)
96      CONTINUE
        DO 98 I=1,NM1
        L=NM-I+1
        NM2=L-1
        DO 98 J=1,NM2
        DO 98 K=1,NM
98      AI(J,K)=AI(J,K)-AI(L,K)*A(J,L)
        WRITE(6,102)((A(I,J),J=1,3),I=1,3),((AI(I,J),J=1,3),I=1,3)
   102  FORMAT(3E20.8/)
        STOP
        END
```

Example Problem   Program P.7

The preceding program is applied to the problem of approximating a semi-circular beam by N straight segments. This problem is useful because the cantilever flexibility is available in Timoshenko (p. 86) for the case when length changes are negligible.

$$C = \begin{bmatrix} \dfrac{\pi}{2}\dfrac{R^3}{EI} & & \text{Sym.} \\[2ex] \dfrac{2R^3}{EI} & \dfrac{3}{2}\dfrac{\pi R^3}{EI} & \\[2ex] \dfrac{2R^2}{EI} & \dfrac{\pi R^2}{EI} & \dfrac{\pi R}{EI} \end{bmatrix}$$

When $R = 240''$, $I = 1728 \ m^4$, $E = 30 \times 10^6$ psi,

$$C = \begin{bmatrix} 0.417 \times 10^{-3} & & \text{Sym.} \\ 0.532 \times 10^{-3} & 0.125 \times 10^{-2} & \\ 0.222 \times 10^{-5} & 0.35 \times 10^{-5} & 0.146 \times 10^{-7} \end{bmatrix}$$

The following program generates input to the program just described which in turn generates the output given.

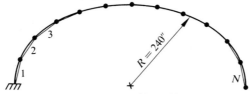

Program P.7. Sample problem.

Fortran Program

```
DIMENSION A(100),TH(100),AI(100),AL(100)
PI=3.14159
R=240.
B=12.
T=12.
N=20
ALP=PI/FLOAT(N)
BETA=PI/2.-ALP/2.
A1=2.*R*SIN(ALP/2.)
R1=2.*R
WRITE(6,101) N,R1
DO 1 I=1,N
A(I)=B*T
TH(I)=BETA
BETA=BETA-ALP
AI(I)=A(I)*T*T/12.
AL(I)=A1
1 WRITE(6,100) A(I),TH(I),AI(I),AL(I)
101 FORMAT(I5,E20.8)
100 FORMAT(4E20.8)
STOP
END
```

Computer Output

$$Case\ 1\ -\ N = 10$$

0.41042408E-03	0.52675675E-03	0.21947981E-05
0.52675745E-03	0.12447410E-02	0.34762725E-05
0.21947972E-05	0.34762679E-05	0.14484382E-07

$$= C_{N=10}$$

0.10000000E 01	0.26737929E-02	-0.26738415E-02
0.10973863E-05	0.10000000E 01	0.44344258E 00
-0.10974109E-05	0.10945369E-08	0.10000000E 01

0.12845281E 05	-0.19464080E 07	0.19464450E 07
-0.19464050E 07	0.50431437E 09	-0.22363866E 09
0.19464490E 07	-0.22363954E 09	0.50432384E 09

$$= K_{N=10}$$

$$Case\ 2\ -\ N = 20$$

0.41681039E-03	0.53167716E-03	0.22153381E-05
0.53167879E-03	0.12536943E-02	0.34869772E-05
0.22153490E-05	0.34869763E-05	0.14529292E-07

$$= C_{N=20}$$

0.10000000E 01	0.26575108E-02	-0.26574670E-02
0.11076854E-05	0.10000000E 01	0.44856715E 00
-0.11076636E-05	0.11204746E-08	0.10000000E 01

0.12653828E 05	-0.19293520E 07	0.19294070E 07
-0.19293670E 07	0.50118144E 09	-0.22482408E 09
0.19294110E 07	-0.22482230E 09	0.50120499E 09

$$= K_{N=20}$$

P.8 PLASTIC ANALYSIS OF PLANE FRAMES

This program is part of the solution of Exercise 4 of Chapter 4; the reader should begin his study of it there. It computes the plastic collapse load of an arbitrary plane frame with fixed supports which is loaded at its joints. It requires as input the ratios of the applied joint loads and the member properties; it computes the plastic collapse load and a set of associated member forces. This program is based upon the lower bound theorem of plastic analysis and uses linear programming as a method of solution.

Introduction

This program contains two steps. The first step sets up the linear programming problem and is largely concerned with generating the equilibrium equations $\tilde{N}F = P$. This step is simply a straightforward modification of Program P.3 and should not be attempted before the reader has familiarized himself with that program. The second step in the program executes the routine MPS (IBM Manual #20-0476-1, "Mathematical Programming System 1360 (360A-CO-14x) Linear and Separable Programming – User's Manual") as a "black box". Indeed, most of the effort of preparing this program is concerned with passing the information generated in the first step to the second with the proper format.

Step by Step Explanation of Program P.8

Since this program is for the most part a modification of Program P.3, only their differences will be emphasized. The reader is again referred to the description of Program P.3.

Step 1   Initialize:
In plastic analysis the yield stress FY is needed rather than Young's modulus. In this first step FY is set to be 36,000 psi (mild steel).

Step 2   Input with echo.

Step 3   Zero the matrix $C = \tilde{N}$.

Step 4   Construct the matrix C by adding the contributions of the individual members. In this program
A(k) – plastic section modulus of member k. (The plastic section modulus is usually referred to as Z and is defined to be μ_k/FY. the moment capacity divided by the yield stress.)

Step 5   Output to MPS system. This section is straightforward and the reader is referred to the appropriate MPS manuals for the required format. The following remarks are intended to simplify this section for the reader:

1. Using the usual linear programming designation of constraints.

$$Ax \leqslant b$$

the matrix x used here is

$$x = \begin{bmatrix} F_1 \\ F_2 \\ \vdots \\ F_B \\ \lambda \end{bmatrix}$$

2. As usual, the first row of the matrix A describes the objective function.
3. The next $(3*(NN - NS))$ rows express equilibrium.
4. The final $(4*NB)$ rows express the requirement that

$$|m_i^+| \leq \mu_i \quad \text{and} \quad |m_i^-| \leq \mu_i$$

This appears as

$$m_i^+ \leq \mu_i \qquad m_i^- \leq \mu_i$$
$$-m_i^+ \leq \mu_i \qquad -m_i^- \leq \mu_i$$

for each member.

Step 6   Execute the MPS program.

Fortran Program

```
//STEP1 EXEC FORTGCLG
//FORT.SYSIN DD *
C      PLASTIC ANALYSIS OF PLANE FRAMES USING MPS
       DIMENSICN A(100),AL(100),              TH(100),LP(100),MI(100)
       DIMENSION P(100),C(100,100),R(3,3),SN(3,3),NRCW(2),RVAL(2)
       R(1,3)=0.
       R(2,3)=0.
       R(3,1)=0.
       R(3,2)=0.
       R(3,3)=1.
       PI=3.14159     /180.
       SN(1,3)=0.
       SN(2,1)=0.
       SN(1,2)=0.
       SN(3,1)=0.
       FY=36000
    1  READ(5,2,END=400)NB,NN,NS
       NNS=NN-NS
    2  FORMAT (3(I3,3X))
    3  FORMAT(3(E10.2,2X),2(3X,I3))
       DO 204 I=1,NNS
       J=3*I-2
  204  READ           (5,3)P(J),P(J+1),P(J+2)
       WRITE(6,903)
       WRITE          (6,902)(I,P(3*I-2),P(3*I-1),P(3*I),I=1,NNS )
  902  FORMAT(I4,3E20.8)
  903  FORMAT(1H1,11HJOINT LOADS /13X,2HPX,18X,2HPY,19X,1HM//)
       WRITE          (6,901)
       N=3*NN
       N3=3*NB
       DO 904 I=1,N3
       DO 904 J=1,N
  904  C(I,J)=0.
       N=3*NNS
       K=0
  926  K=K+1
       READ           (5,3)A(K),AL(K),      TH(K),LP(K),MI(K)
       WRITE          (6,900)K, A(K),AL(K),       TH(K),LP(K),MI(K)
  900  FORMAT(I4,3E20.8,2I10)
  901  FORMAT(18H1MEMBER PROPERTIES /14X,1HZ   ,15X,6HLENGTH,       17X,
      1 5HANGLE,14X,5H+ END,5X,5H- END//)
       ZK=1.
       II=3*LP(K)-3
       J=1
   12  GO TO (13,11,      26),J
   11  II=3*MI(K)-3
       ZK=-1.
       GO TO 352
   13  ANG=TH(K)*PI
       R(1,1)= COS(ANG)
       R(2,2)=R(1,1)
       R(1,2)= SIN(ANG)
       R(2,1)=-R(1,2)
       Q=0.
  352  SN(1,1)=ZK
       SN(2,2)=-ZK/AL(K)
       SN(3,2)=-ZK/AL(K)
       IF(ZK) 15,15,16
   16  SN(2,3)=1.
       SN(3,3)=0.
```

INITIALIZE ARRAYS

INPUT & ECHO

CONSTRUCT THE MATRIX

SUBROUTINE FOR R_i & $\bar{R}_i$

```
      GO TO 24
   15 SN(2,3)=0.
      SN(3,3)=1.
   24 DO 27 I1=1,3
      N3=3*K-3+I1
      DO 27 I2=1,3
      N4=II+I2
      DO 27 I3=1,3
   27 C(N3,N4)=C(N3,N4)+          SN(I1,I3)*R(I3,I2)
      J=J+1
      GO TO 12
   26 IF(K-NB)926,927,927
  927 REWIND 7
      WRITE(7,401)
  401 FORMAT(4HNAME,10X,5HFRAME/4HROWS/1X,1HN,2X,5HWEIGH)
      DO 402 I=1,N
      J=I+100
  402 WRITE(7,403)J
  403 FORMAT((5H E  R,I3))
      I1=4*NB
      DO 404 I=1,I1
      J=N+100+I
  404 WRITE(7,405)J
  405 FORMAT((5H L  R,I3))
      WRITE(7,406)
  406 FORMAT(7HCOLUMNS)
      NBR=N
      N3=3*NB
      DO 407 I=1,N3
      I1=1
      DO 408 J=1,N
      IF(C(I,J))409,408,409
  409 NROW(I1)=J
      RVAL(I1)=C(I,J)
      CALL WRITE(I1,I,NROW,RVAL)
  408 CONTINUE
      IF(MOD(I,3)-1)414,413,414
  414 NROW(I1)=NBR+1
      RVAL(I1)=1.
      CALL WRITE(I1,I,NROW,RVAL)
      NROW(I1)=NBR+2
      RVAL(I1)=-1.
      CALL WRITE(I1,I,NROW,RVAL)
      NBR=NBR+2
  413 IF(I1-1) 417,407,417
  417 NROW(I1)=NBR-3
      RVAL(I1)=0.
      CALL WRITE(I1,I,NROW,RVAL)
  407 CONTINUE
      K=N3+101
      WRITE(7,431)K
  431 FORMAT(4X,1HC,I3,6X,5HWEIGH,5X,2H1.)
      I=N3+1
      I1=1
      DO 430 M=1,N
      IF(P(M))432,430,432
  432 NROW(I1)=M
      RVAL(I1)=-P(M)
      CALL WRITE(I1,I,NROW,RVAL)
  430 CONTINUE
```

GENERATE INPUT FOR MPS PROGRAM

```
      IF(I1-1) 434,433,434
434   NROW(I1)=N+1
      RVAL(I1)=0.
      CALL WRITE(I1,I,NROW,RVAL)
433   WRITE(7,422)
422   FORMAT(3HRHS)
      K=N+100
      DO 421 I=1,NB
      DO 421 J=1,4
      K=K+1
      AB=A(I)*FY
421   WRITE(7,423)K,AB
423   FORMAT(4X,5HRHS01,5X,1HR,I3,6X,G12.6)
      WRITE(7,424)
424   FORMAT(6HBOUNDS)
      DO 425 I=1,N3
      K=I+100
425   WRITE(7,426)K
426   FORMAT(9H FR FORCE,5X,1HC,I3)
      WRITE(7,427)
427   FORMAT(6HENDATA/2H/*)
      GO TO 1
400   RETURN
      END
      SUBROUTINE WRITE(I1,I,NROW,RVAL)
      DIMENSION NROW(2),RVAL(2)
      I1=I1+1
      IF(I1-3)408,410,408
410   I1=1
      K=100+I
      K1=NROW(1)+100
      K2=NROW(2)+100
      WRITE(7,411) K,K1,RVAL(1),K2,RVAL(2)
411   FORMAT(4X,1HC,I3,6X,1HR,I3,6X,G12.5,3X,1HR,I3,6X,G12.5)
408   RETURN
      END
/*
//GO.FT07F.001 DD DISP=(NEW,PASS),UNIT=SYSDA,SPACE=(80,(200,10)),
//              DCB=(RECFM=F,BLKSIZE=80,LRECL=80)
//GO.SYSIN DD *
    4     5     2
1.
0.
0.
        100.        120.        90.      1      4
        69.         169.        45.      2      1
        69.         169.        -45.     3      2
        100.        120.        90.      3      5
/*
//STEP2 EXEC MPS
//COMPIL.SYSIN DD    *
        PROGRAM
        INITIALZ
        MOVE(XDATA,'FRAME')
        MOVE(XPBNAME,'PBFILE')
        CONVERT
        SETUP('BOUND','FORCE','MAX')
        MOVE(XOBJ,'WEIGH')
        MOVE(XRHS,'RHS01')
        PRIMAL
```

SUBROUTINE FOR WRITING

DATA FOR EXAMPLE

MPS ROUTINE

```
        SCLUTION
        PICTURE
        EXIT
        PEND
/*
//EXECUT.SYSIN   DD  DSN=*.STEP1.GO.FT07F001.DISP=(SHR,DELETE)
/*
```

Sample Problem

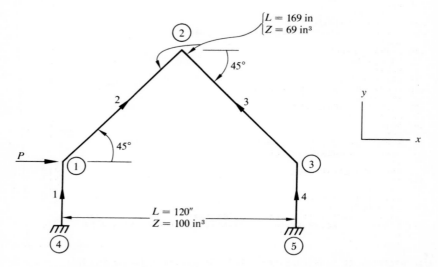

$$\begin{cases} L = 169 \text{ in} \\ Z = 69 \text{ in}^3 \end{cases}$$

45°

45°

P

$$L = 120''$$
$$Z = 100 \text{ in}^3$$

y

x

Program P.8. Sample problem. Find the plastic collapse load P.

Partial Computer Output

```
                EXECUTOR.     MPS/360 V2-M8

    SECTION 2 - CCLUMNS
```

NUMBER	.COLUMN.	AT	...ACTIVITY...	..INPUT COST..	..LOWER LIMIT.	..UPPER L
27	C101	BS	10787.86613	.		NONE
28	C102	BS	2483999.99998	.		NONE
29	C103	BS	3600000.00000	.		NONE
30	C104	BS	14140.45240-	.		NONE
31	C105	BS	2483999.99998-	.		NONE
32	C106	BS	2483999.99998-	.		NONE
33	C107	BS	29396.57026-	.		NONE
34	C108	BS	94273.34459-	.		NONE
35	C109	BS	2483999.99998	.		NONE
36	C110	BS	10787.68511-	.		NONE
37	C111	BS	94273.34459	.		NONE
38	C112	BS	3600000.00000	.		NONE
39	C113	BS	81485.28526	1.00000		

168

Index